T0275691

Modeling and Analysis of Doubly Fed Induction Generator Wind Energy Systems

Modeling and Analysis of Doubly Fed Induction Generator Wind Energy Systems

Lingling Fan and Zhixin Miao

AMSTERDAM • BOSTON • HEIDELBERG • LONDON
NEW YORK • OXFORD • PARIS • SAN DIEGO
SAN FRANCISCO • SINGAPORE • SYDNEY • TOKYO
Academic Press is an imprint of Elsevier

Academic Press is an imprint of Elsevier
125 London Wall, London, EC2Y 5AS, UK
525 B Street, Suite 1800, San Diego, CA 92101-4495, USA
225 Wyman Street, Waltham, MA 02451, USA
The Boulevard, Langford Lane, Kidlington, Oxford OX5 1GB, UK

Notices
Knowledge and best practice in this field are constantly changing. As new research and
experience broaden our understanding, changes in research methods, professional practices,
or medical treatment may become necessary.

Practitioners and researchers must always rely on their own experience and knowledge in
evaluating and using any information, methods, compounds, or experiments described herein.
In using such information or methods they should be mindful of their own safety and the safety
of others, including parties for whom they have a professional responsibility.

To the fullest extent of the law, neither the Publisher nor the authors, contributors, or editors,
assume any liability for any injury and/or damage to persons or property as a matter of products
liability, negligence or otherwise, or from any use or operation of any methods, products,
instructions, or ideas contained in the material herein.

British Library Cataloguing in Publication Data
A catalogue record for this book is available from the British Library

Library of Congress Cataloging-in-Publication Data
A catalog record for this book is available from the Library of Congress

ISBN: 978-0-12-802969-5

For information on all Academic Press publications
visit our website at http://store.elsevier.com/

Working together
to grow libraries in
developing countries

ELSEVIER   Book Aid International

www.elsevier.com • www.bookaid.org

DEDICATION

To our parents

CONTENTS

Acknowledgments viii

Chapter 1 Introduction 1
1.1 Wind Energy Integration Issues 1
1.2 Objectives of This Book 3
1.3 Structure of the Book 4
References 6

Chapter 2 AC Machine Modeling 8
2.1 Space Vector and Complex Vector Explanation 8
2.2 Derivation of an Induction Machine Modeling in Space Vector
and Complex Vector 13
2.3 DFIG Modeling 21
2.4 Examples 23
References 33

Chapter 3 Modeling of Doubly Fed Induction Generation (DFIG)
Converter Controls 34
3.1 Rotor Flux-Oriented Induction Machine Control 35
3.2 DFIG Rotor Side Converter Control 40
3.3 GSC Control 46
3.4 Complete DFIG Modeling Blocks 48
3.5 Examples 49
References 54

Chapter 4 Analysis of DFIG with Unbalanced Stator Voltage 55
4.1 Steady-State Harmonic Analysis of a DFIG 55
4.2 Unbalanced Stator Voltage Drop Transient Analysis 64

4.3   Converter Control to Mitigate Unbalance Effect 67

References 73

Chapter 5 State-Space Based DFIG Wind Energy
        System Modeling 74

5.1   State-Space Model of a Series
      Compensated Network 75

5.2   State-Space Model of DFIG Wind Energy System.................... 76

5.3   Integrated System Model 78

5.4   Application of SSR Analysis 80

Appendix.. 92

References 93

Chapter 6 Frequency-Domain Based DFIG Wind Energy Systems
        Modeling 94

6.1   Introduction of Impedance Model and Its Application in Stability
      and Resonance Detection.. 95

6.2   Impedance Models of DFIG in Various
      Reference Frames 96

6.3   Negative-Sequence DFIG Impedance Model.......................... 103

6.4   Inclusion of DFIG into Torque-Speed Transfer Functions 106

6.5   Examples 111

References 127

Chapter 7 Multi-Machine Modeling and Inter-Area Oscillation
        Damping 128

7.1   Steady-State Calculation for a DFIG 129

7.2   Interconnection of DFIG Model in Power Systems.................... 132

7.3   Application of the Model: Interarea-Oscillation Damping
      Through DFIG Wind Turbines 133

Appendix.. 144

References 145

ACKNOWLEDGMENTS

We are fortunate to have the opportunity and time to write this monograph. Through our career development, we are blessed with the encouragement and support from mentors, colleagues, and friends. Our thanks go to Naihu Li, Guangxiang Lu, Haifeng Wang, Xiao-Ping Zhang, Mariesa Crow, Yilu Liu, Elham Makram, Dagmar Niebur, Fang Z. Peng, Jian Sun, and Kevin Tomsovic.

Our research in wind energy grid integration started from our years at Midwest ISO and spanned through time spent at North Dakota State University. We would like to acknowledge our mentors and collaborators at MAPPCOR, MISO, and North Dakota State University: Dale Osborn, Jeff Webb, Larry Brusseau, Terry Bilke, Jerald Miland, Subbaraya Yuvarajan, and Rajesh Kavasseri. Their help is enormous.

The research in wind energy grid integration is closely related to power electronics, a fairly new area to us when we started our investigation. There are many outstanding researchers in the field whose papers we digested and benefited a great deal. Profs. Jian Sun and Fang Z. Peng are sources of inspiration.

Our thanks also go to the former and current graduate students at University of South Florida Smart Grid Power Systems Lab. Their contributions and team spirit make USF Smart Grid Power Systems Lab a great place to work at. Haiping Yin's work in DFIG modeling and unbalanced mitigation control contributed to this book. Javad Khazaei developed a few examples in Matlab/SimPowerSystems for this book. Ling Xu, Lakshan Piyasinghe, and Javad Khazaei have extended the modeling techniques to other systems.

Our thanks go to Bikash Pal. Without his recommendation to Elsevier, this book would not be possible.

CHAPTER 1

Introduction

1.1 WIND ENERGY INTEGRATION ISSUES 1

1.2 OBJECTIVES OF THIS BOOK 3

1.3 STRUCTURE OF THE BOOK 4

REFERENCES 6

1.1 WIND ENERGY INTEGRATION ISSUES

Integration issues related to wind energy can be roughly categorized into two types based on time scales: operation issues and dynamic issues. Operation issues relate to a long-term time scale, e.g., hours, days, or weeks. Operating the system and dispatching synchronous generators while accommodating intermittent wind energy at hourly base or daily base is an example. For this type of research, wind farms' generation profiles of hours and days are of study interest. This type of problems belongs to the category of operations research. Optimization problems are usually formulated and solved. Example research can be found in [1–3]. In [1], a 24-hour unit commitment problem based on hourly generation, load, and wind profiles is formulated to evaluate the impact of wind penetration on market price. An economic dispatch problem with wind energy is presented in [2], where the stochastic nature of wind power is modeled based on wind speed Weibull probability distribution function. In [3], a short-term forward electricity market clearing problem is formulated with uncertain wind.

Compared to operation issues, dynamic issues are of very short time scales. For example, the time-scale of inter-area oscillations and electrome-chanical resonances is tens of seconds. The time scale of electric resonances is be even shorter.

Based on the above classification, many issues related to wind energy grid integration are dynamic issues, e.g., low voltage ride through, sub-synchronous resonances (SSR), frequency support from wind, and control coordination of wind and high voltage direct current (HVDC) delivery

Modeling and Analysis of Doubly Fed Induction Generator Wind Energy Systems.
http://dx.doi.org/10.1016/B978-0-12-802969-5.00001-9

systems. Dynamic modeling of wind energy system is important to approach this type of research. For example, in [4, 5], dynamic models of an induction machine are used to investigate the effect of stator voltage drop on transients in rotor voltages and currents. In our research on SSR in Type-3 wind farms with series-compensated networks [6], dynamic models of a Type-3 wind energy system and the grid were built. Analysis and simulation were conducted based on this model. In [7], frequency support schemes from wind energy system are validated through dynamic model-based simulation in the time scale of tens of seconds. Dynamic model-based simulation is also employed when designing DFIG (doubly fed induction generator) and HVDC control coordination [8].

To be able to investigate the aforementioned issues, we need a good understanding of the dynamic models of wind energy systems as well as the grid. Compared to conventional power systems where synchronous generators are the main source, wind energy systems include both ro-tating machines and power converter interfaces. Thus, modeling has to capture power converter control dynamics as well as the electromagnetic and electromechanical dynamics of AC machines. In conventional power systems, electromechanical dynamics (e.g., low-frequency oscillations) and torsional interactions between the turbine mechanical systems and the elec-tric systems are expected. In wind energy systems, inclusion of converters could introduce resonances due to converter-grid interactions and machine-converter interactions. Investigation of emerging dynamic phenomena re-quires adequate dynamic models of wind energy systems.

Wind turbines are classified into four types based on AC generator type and converter type [9]. Type-1 systems deploy induction machines and work at a fixed wind speed. Type-2 wind turbines use wound rotor induction generators with rotor resistance control. Type-3 and Type-4 wind turbines are called variable speed fixed frequency wind generators. These generators can work at variable wind speed and produce fixed frequency electricity. Type-3 wind generator uses doubly-fed induction generator (DFIG) with a partial converter interface, which connects the rotor side to the grid through a back-to-back converter (a rotor-side converter (RSC) and a grid-side converter (GSC) connected via a DC-link capacitor), as shown in Fig. 1.1. Type-4 wind generator uses permanent magnet synchronous generator with a full-converter interface.

The state-of-the-art wind energy systems adopt voltage source converter-enabled wind turbines: Type-3 and Type-4. Type-3 occupies more market

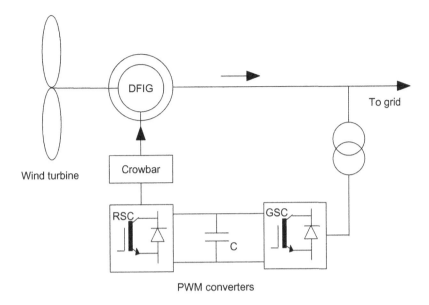

Figure 1.1 Type-3 wind energy system.

than Type-4. GE Wind's 1.5/1.6 MW wind turbine (Type-3) was the most popular wind turbine in the U.S. market in 2012 [10].

The focus of this book is Type-3 wind energy systems. Dynamic modeling of DFIG-based wind energy system and its applications in dynamic studies will be elaborated in this book.

1.2 OBJECTIVES OF THIS BOOK

With the focus on Type-3 or DFIG wind energy systems, throughout this book, analytical DFIG models will be derived and applied in various studies. Currently, there is a lack of monographs on systematic treatment of DFIG wind energy system in terms of modeling, analysis, and control.

There are classic textbooks on modeling and analysis of AC machines and drives, e.g., Krause's *Analysis of Electric Machinery* and Bose's *Power Electronics and AC Drives*. These books focus on drive systems. There are significant differences in control objectives for an AC drive system and a grid integration system. In drive systems, machine control focuses on flux and torque control, while grid integration focuses on real and reactive power control. In addition, many grid integration dynamic phenomena are system-level phenomena, e.g., SSR. System models including wind generation are required. These phenomena emerged and attracted interest from the research community and industry. These topics were not covered in a drive book.

The purpose of writing this book is to integrate the state-of-the-art wind energy grid integration research into education. The book can serve as a textbook for a graduate-level course on AC machine modeling with applications in wind energy integration. The book can also be read by researchers and practitioners in the field of wind energy integration.

1.3 STRUCTURE OF THE BOOK

The book is organized as follows.

Chapter 2 discusses AC machine modeling. Class notes from the authors' course on "AC Machines and Drives" at University of South Florida will be used to develop AC machine modeling using space vectors and complex vectors. Following induction machine modeling, DFIG machine modeling will be discussed. Simulation and analysis examples will be provided.

Chapter 3 presents modeling of DFIG converter control. In this chapter, converter control of DFIG will be thoroughly explained. Differences between drive control and grid integration control will be discussed. In this chapter, a simulation example in PSCAD with power electronic switching details will be given. The purpose of the demonstration is to show that converter voltage outputs after filters are sinusoidal. Readers can then build visual connections between control of abc sinusoidal waveforms versus vector control or control of dq-variables. Modeling of converter controls of DFIG will then be explained based on dq-reference frame. The mathematical models will be derived and a simulation example will be demonstrated.

Chapter 4 discusses unbalanced DFIG analysis and control. In this chapter, steady-state analysis and transient analysis of DFIG rotor voltage/current due to unbalanced stator voltage drop will be presented. Following the analysis, mitigation of unbalance effect using converter control is presented. Simulation case studies are presented to verify the mitigation strategies.

Chapter 5 presents state-space-based DFIG wind energy system modeling. The state-space-based DFIG model in the dq-reference frame will be presented. Integrated system model of a DFIG with a series-compensated transmission line is presented next. The developed model will be used for SSR analysis. In this chapter, our research on DFIG-related SSR [6, 11, 12] will be elaborated in both modeling and applications of the dynamic models for small-signal analysis as well as dynamic simulation.

Chapter 6 presents frequency-domain-based DFIG wind energy systems modeling. DFIG's frequency-domain models at phase domain (positive and negative sequences) will be derived and explained.

Three examples will be used to explain the usage of frequency-domain-based DFIG models. The first example examines SSR detection and considers only balanced operating conditions. The second example considers unbalanced operation conditions and the effect on SSR. The third example uses torque-speed transfer function to explain DFIG's influence on torsional interactions. Our research work in [13–15] will be elaborated in this chapter.

Chapter 7 presents multimachine modeling and inter-area oscillation damping. In this chapter, steady-state calculating of DFIG is first presented. Then interfacing a DFIG dynamic model in the dq reference frame with a multi-machine power system is explained. A design example on inter-area oscillation damping is given to demonstrate the use of such a model.

The flow of information of this book can be visually seen from Fig. 1.2. Chapters 2 and 3 deal with AC machine and converter control, respectively. With both machine and power converter interface modeled, a DFIG wind energy system model is possible. Chapter 4 can be considered as an application of the system model in unbalanced analysis and control. Note that up to Chapter 4, we only consider a simple system where a wind energy system is directly connected to a grid. The transmission line modeled as an resistor-inductor or RL circuit can be considered as the additional stator resistance and leakage inductance. The dynamic model will have a given stator voltage to start with.

Following Chapter 4, we now deal with more complicated transmission systems, first a series-compensated transmission line where the dynamics of the resistor-inductor-capacitor (RLC) line should not be ignored (Chapters 5 and 6), and then a multi-machine power system (Chapter 7). The interfacing techniques between wind generation and grid are detailedly explained in Chapter 5 and Chapter 7. Chapter 6 can be considered as parallel to Chapter 5. In addition to state-space modeling, we also look for another type of modeling: impedance modeling. This frequency-domain modeling approach shows its advantage in modularized modeling and its easiness in handling unbalanced conditions.

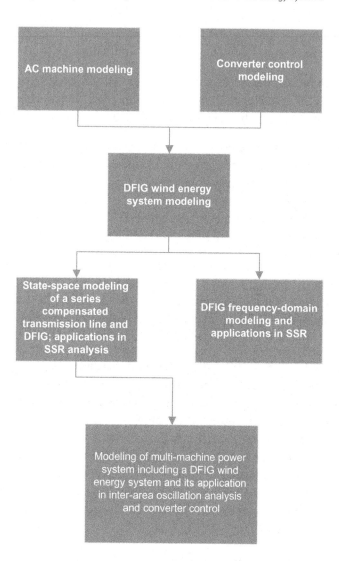

Figure 1.2 Information flow of this book.

REFERENCES

[1] J. Garcia-Gonzalez, R.R. de la Muela, L.M. Santos, A.M. González, Stochastic joint optimization of wind generation and pumped-storage units in an electricity market, IEEE Trans. Power Syst. 23(2) (2008) 460-468.

[2] J. Hetzer, D.C. Yu, K. Bhattarai, An economic dispatch model incorporating wind power, IEEE Trans. Energy Convers. 23(2) (2008) 603-611.

[3] F. Bouffard, F.D. Galiana, Stochastic security for operations planning with significant wind power generation, in: 2008 IEEE Power and Energy Society General Meeting—Conversion and Delivery of Electrical Energy in the 21st Century, IEEE, 2008, pp. 1-11.

[4] J. Lopez, P. Sanchis, X. Roboam, L. Marroyo, Dynamic behavior of the doubly fed induction generator during three-phase voltage dips, IEEE Trans. Energy Convers. 22(3) (2007) 709-717.

[5] J. Lopez, E. Gubia, P. Sanchis, X. Roboam, L. Marroyo, Wind turbines based on doubly fed induction generator under asymmetrical voltage dips, IEEE Trans. Energy Convers. 23(1) (2008) 321-330.

[6] L. Fan, R. Kavasseri, Z.L. Miao, C. Zhu, Modeling of DFIG-based wind farms for SSR analysis, IEEE Trans. Power Delivery 25(4) (2010) 2073-2082.

[7] J. Morren, S.W. De Haan, W.L. Kling, J. Ferreira, Wind turbines emulating inertia and supporting primary frequency control, IEEE Trans. Power Syst. 21(1) (2006) 433-434.

[8] H. Yin, L. Fan, Z. Miao, Fast power routing through HVDC, IEEE Trans. Power Delivery 27(3) (2012) 1432-1441.

[9] Z. Chen, J.M. Guerrero, F. Blaabjerg, A review of the state of the art of power electronics for wind turbines, IEEE Trans. Power Electron. 24(8) (2009) 1859-1875.

[10] R. Wiser, 2012 wind Technologies Market Report, 2014. http://www1.eere.energy.gov/wind/pdfs/2012_wind_technologies_market_report.pdf.

[11] L. Fan, C. Zhu, Z. Miao, M. Hu, Modal analysis of a DFIG-based wind farm interfaced with a series compensated network, IEEE Trans. Energy Convers. 26(4) (2011) 1010-1020.

[12] L. Fan, Z. Miao, Mitigating SSR using DFIG-based wind generation, IEEE Trans. Sustainable Energy 3(3) (2012) 349-358.

[13] L. Fan, Z. Miao, Nyquist-stability-criterion-based SSR explanation for type-3 wind generators, IEEE Trans. Energy Convers. 27(3) (2012) 807-809.

[14] Z. Miao, Impedance-model-based SSR analysis for type 3 wind generator and series-compensated network, IEEE Trans. Energy Convers. 27(4) (2012) 984-991.

[15] Z. Miao, Impact of unbalance on electrical and torsional resonances in power electronic interfaced wind energy systems, IEEE Trans. Power Syst. 28(3) (2013) 3105-3113.

CHAPTER 2

AC Machine Modeling

2.1 SPACE VECTOR AND COMPLEX VECTOR EXPLANATION 8

2.1.1 Examples of Space Vector 12

2.2 DERIVATION OF AN INDUCTION MACHINE MODELING IN
   SPACE VECTOR AND COMPLEX VECTOR 13

2.2.1 Induction Machine Modeling in Space Vector 13

2.2.1.1 Per Unit System 15

2.2.2 Induction Machine Modeling in Complex Vector 16

2.2.2.1 Model in Per-Unit System 18

2.2.3 Swing Equation 18

2.3 DFIG MODELING 21

2.3.1 Wind Turbine Aerodynamics Model 21

2.4 EXAMPLES 23

2.4.1 Example 1: Free Acceleration 23

2.4.1.1 Induction Machine Simulink Model 25

2.4.2 Example 2: DFIG Stator Voltage Drop 27

REFERENCES 33

In this chapter, analytical models of induction machines are first described. Analytical models of DFIG are then presented. The models are related to electric machine only. Converter controls will be discussed in Chapter 3. Two examples are given to demonstrate the usage of the analytical models to simulate free acceleration of an induction machine or analyze the consequences of DFIG stator voltage dip.

2.1 SPACE VECTOR AND COMPLEX VECTOR EXPLANATION

The type of induction machines to be discussed is three-phase induction machines where both the stator and the rotor have *abc* windings. The key dynamics of the magnetic field can be described by Faraday's law that the change of flux field induces electromagnetomotive force (EMF). To model a

Modeling and Analysis of Doubly Fed Induction Generator Wind Energy Systems.
http://dx.doi.org/10.1016/B978-0-12-802969-5.00002-0

three-phase system, a direct approach is to express the EMF phase by phase. However, this approach will lead to the use of phase-coupled and time-varied inductances. The contribution of Park's transformation is to develop analytical models based on a *dq* rotating reference frame. When the rotating speed of the reference frame is the same as the speed of the rotating flux, the models will have a much simpler form and the inductances of the *dq*-axis are decoupled and constant.

In this chapter, however, we will not use Park's transformation to derive the analytical models. Instead, we rely on the concept of space vector to derive the analytical models due to its insightful and simple explanation of physics. In this section, let us first define and explain space vector as well as complex vector.

With the assumptions such as sinusoidal distribution of air gap flux and a uniform air gap, three-phase balanced stator currents in a nonsalient two-pole AC machine (Fig. 2.1) will lead to an magnetomotive force (MMF) which can be viewed as a traveling waveform with a constant magnitude in the air gap (Fig. 2.2) or a rotating MMF—a space vector. This MMF leads to a rotating flux in the air gap (Fig. 2.1). The rotating flux in the air gap then induces sinusoidal EMFs in the three-phase stator and rotor windings.

The concept of space vector comes from this physical mechanism: three-phase currents lead to a rotating MMF. For each phase current, at a random position (defined as angle α referring to the *a*-axis) in the air gap, the resulting MMF can be found based on Ampere's law. The mathematical expressions are listed as follows:

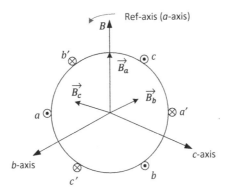

Figure 2.1 A three-phase AC machine.

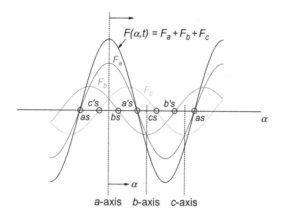

Figure 2.2 MMF distribution in the air gap.

$$F_a(\alpha) = Ni_a \cos \alpha$$

$$F_b(\alpha) = Ni_b \cos \left(\alpha - \frac{2\pi}{3} \right)$$

$$F_c(\alpha) = Ni_c \cos \left(\alpha + \frac{2\pi}{3} \right)$$

where α is the general angle in the air gap referring to the a-axis, F is MMF and N is the number of windings in each phase.

From the above equations, we can find that F_a reaches maximum when $\alpha = 0$. Accordingly, F_b is maximum when $\alpha = \frac{2\pi}{3}$; F_c is maximum when $\alpha = \frac{4\pi}{3}$. Consider the current i as a sinusoidal function of time t:

$$i_a(t) = I_m \cos(\omega_s t + \theta_a)$$

$$i_b(t) = I_m \cos(\omega_s t + \theta_a - \frac{2\pi}{3})$$

$$i_c(t) = I_m \cos(\omega_s t + \theta_a + \frac{2\pi}{3})$$

where I_m is the amplitude of the current, ω_s is the angular frequency of the current and θ_a is the initial angle at $t = 0$.

Then the total MMF can be found as

$$F(\alpha, t) = NI_\mathrm{m} \left[\cos(\omega_s t + \theta_a) \cos\alpha + \cos\left(\omega_s t + \theta_a - \frac{2\pi}{3}\right) \cos\left(\alpha - \frac{2\pi}{3}\right)\right.$$
$$\left. + \cos\left(\omega_s t + \theta_a + \frac{2\pi}{3}\right) \cos\left(\alpha + \frac{2\pi}{3}\right)\right]$$
$$= \frac{3}{2} NI_\mathrm{m} \cos(\alpha - \omega_s t - \theta_a) \tag{2.1}$$

Remarks. It is obvious that the total MMF due to a set of three-phase balanced currents in a nonsalient AC machine has a constant magnitude. The total MMF in the air gap is a traveling waveform. As time evolves, the peak of the waveform moves forward. Figure 2.2 illustrates the MMFs due to per-phase currents and the total MMF at an instant in the air gap. This MMF can be viewed as a space vector notated by a magnitude and an angle where the peak occurs. The MMF can be notated as

$$\overrightarrow{F}(t) = \frac{3}{2} NI_\mathrm{m} e^{j(\omega_s t + \theta_a)} = \frac{3}{2} NI_\mathrm{m} e^{j\theta_a} e^{j\omega_s t} \tag{2.2}$$

It can be observed that three-phase balanced currents result in a rotating MMF space vector with a constant magnitude and a constant speed. Based on the physical mechanism, we can now define a space vector through a similar procedure.

The space vector is defined as

$$\overrightarrow{f}(t) = \frac{2}{3} \left[e^{j0} f_a(t) + e^{j\frac{2\pi}{3}} f_b(t) + e^{j\frac{4\pi}{3}} f_c(t) \right] \tag{2.3}$$

Should $f_a(t)$, $f_b(t)$, and $f_c(t)$ be balanced sinusoidal waveforms with an angular speed ω and initial angle θ_0, and an amplitude of f_m, the space vector can be further expressed as follows:

$$\overrightarrow{f}(t) = |F| e^{j(\omega t + \theta_0)} = |F| e^{j\theta_0} e^{j\omega t} = f_\alpha - j f_\beta \tag{2.4}$$

where $|F| = f_\mathrm{m}$ (the amplitude of the waveforms).

The space vector can be further expressed in real and imaginary components f_α and f_β, where

$$\begin{bmatrix} f_\alpha \\ f_\beta \end{bmatrix} = \frac{2}{3} \begin{bmatrix} 1 & \cos\frac{2\pi}{3} & \cos\frac{2\pi}{3} \\ 0 & -\sin\frac{2\pi}{3} & \sin\frac{2\pi}{3} \end{bmatrix} \begin{bmatrix} f_a \\ f_b \\ f_c \end{bmatrix}$$

2.1.1 Examples of Space Vector

Example 1. Find the space vector for a three-phase voltage expressed as follows:

$$\begin{cases} v_a(t) = \hat{v}_p \cos(\omega t + \theta_p) + \hat{v}_n \cos(\omega t + \theta_n) + \hat{v}_0 \cos(\omega t + \theta_0) \\ v_b(t) = \hat{v}_p \cos(\omega t + \theta_p - \frac{2\pi}{3}) + \hat{v}_n \cos(\omega t + \theta_n + \frac{2\pi}{3}) + \hat{v}_0 \cos(\omega t + \theta_0) \\ v_c(t) = \hat{v}_p \cos(\omega t + \theta_p + \frac{2\pi}{3}) + \hat{v}_n \cos(\omega t + \theta_n - \frac{2\pi}{3}) + \hat{v}_0 \cos(\omega t + \theta_0) \end{cases}$$
(2.5)

Solution. The three-phase voltage consists of positive-, negative-, and zero-sequence components. Based on the definition of space vector, we can find the space vectors corresponding to the positive-, negative-, and zero-sequence components, respectively.

$$\begin{cases} \overrightarrow{v_p} = \hat{v}_p e^{j\theta_p} e^{j\omega t} \\ \overrightarrow{v_n} = \hat{v}_n e^{-j\theta_n} e^{-j\omega t} \\ \overrightarrow{v_0} = 0 \end{cases}$$
(2.6)

The resulting space vector is

$$\overrightarrow{v} = \hat{v}_p e^{j\theta_p} e^{j\omega t} + \hat{v}_n e^{-j\theta_n} e^{-j\omega t}.$$
(2.7)

Note that the positive-sequence component results in a counterclockwise rotating space vector, the negative-sequence component results in a clockwise rotating space vector, and the zero-sequence component results in zero space vector. This fact corroborates with the physics of rotating magnetic field: (1) a set of positive-sequence stator currents result in a counter clockwise rotating magnetic field, (2) a set of negative-sequence stator currents result in a clockwise rotating magnetic field, and (3) a set of zero-sequence stator currents result in no magnetic field.

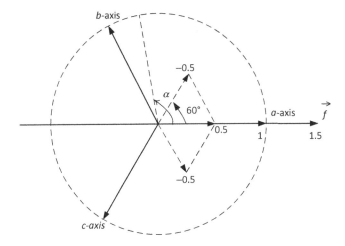

Figure 2.3 Example 1: Space vector illustration.

Example 2. Assume that the stator currents have an angular speed ω_e. When $t = t_1$,

$$\begin{cases} i_a(t_1) = 1 \\ i_b(t_1) = -0.5 \\ i_c(t_1) = -0.5 \end{cases} \tag{2.8}$$

Find the $F(\alpha, t)$.

Solution. We will use a space vector phasor diagram in Fig. 2.3 to find the answer.

Plot 1 in a-axis. Plot -0.5 in b-axis. Plot -0.5 in c-axis. The three vectors add up to 1.5 in a-axis. Therefore, the resulting MMF is

$$\overrightarrow{F}(t) = 1.5 N e^{j\omega_e(t-t_1)} \tag{2.9}$$

$$F(\alpha, t) = 1.5 N \cos(\alpha - \omega_e(t - t_1)) \tag{2.10}$$

2.2 DERIVATION OF AN INDUCTION MACHINE MODELING IN SPACE VECTOR AND COMPLEX VECTOR

2.2.1 Induction Machine Modeling in Space Vector

Through space vector, three-phase instantaneous variables are now integrated into one space vector variable. More important, there is no need to consider the coupling flux linkages among phases. The MMF (flux) space vector is linearly related to the current space vector under the assumption of unsaturated magnetic field. This can greatly simplify the modeling process. In addition, since a space vector can be considered as a weighted sum of three-phase instantaneous variables, the relationship of the instantaneous variables should be true of that of the corresponding space vectors.

Therefore, for the EMF induced in the stator windings, we can write the following equation:

$$\vec{v_s} = R_s \vec{i_s} + \frac{d\vec{\psi_s}}{dt} \tag{2.11}$$

where the subscript s denotes stator-related variables, ψ_s is the flux linkage-linked stator windings. The reference frame is the stationary reference frame for the instantaneous stator winding currents and voltages.

Similarly, for the EMF induced in the rotor windings, we can write the following equation:

$$\vec{v_r^r} = R_r \vec{i_r^r} + \frac{d\vec{\psi_r^r}}{dt} \tag{2.12}$$

where the subscript r denotes rotor-related variables, the superscript r denotes the reference frame is aligned with the rotor, ψ_r^r is the flux linkage linked with the rotor windings observed in the rotor reference frame.

Equations (2.11) and (2.12) are in different reference frames. A same reference frame is sought. Here we choose the stationary reference frame. Assume that the rotating speed of the rotor is ω_m, then the relationship of the rotor flux linkages (rotor currents and EMFs) viewed from the stationary reference frame and the rotor reference frame is as follows:

$$\vec{\psi_r} = \vec{\psi_r^r} e^{j\omega_m t} \tag{2.13}$$

$$\vec{i_r} = \vec{i_r^r} e^{j\omega_m t} \tag{2.14}$$

$$\vec{v_r} = \vec{v_r^r} e^{j\omega_m t} \tag{2.15}$$

Expressing (2.12) in stationary reference frame leads to

$$\vec{v_r} e^{-j\omega_m t} = R_r \vec{i_r} e^{-j\omega_m t} + \frac{d}{dt}(\vec{\psi_r} e^{-j\omega_m t}) \tag{2.16}$$

$$= R_r \vec{i_r} e^{-j\omega_m t} + \left(\frac{d}{dt}\vec{\psi_r}\right) e^{-j\omega_m t} - j\omega_m \vec{\psi_r} e^{-j\omega_m t} \tag{2.17}$$

The concise form of the above equation is as follows:

$$\vec{v_r} = R_r \vec{i_r} + \frac{\mathrm{d}}{\mathrm{d}t}\vec{\psi_r} - j\omega_m \vec{\psi_r} \tag{2.18}$$

To find out the relationship of the stator flux linkage and the rotor flux linkage, we make a few assumptions. We can either assume the windings of the rotor and the stator have the same number, or we can assume that the rotor flux linkage, the rotor current, and the EMF in the rotor circuits are already referred to the stator side by scaling with appropriate turns ratios. Under the above assumptions, we have

$$\vec{\psi_s} = \vec{\psi_m} + \vec{\psi_{ls}} \tag{2.19}$$

$$\vec{\psi_r} = \vec{\psi_m} + \vec{\psi_{lr}} \tag{2.20}$$

$$\vec{\psi_m} = L_m(\vec{i_s} + \vec{i_r}) \tag{2.21}$$

$$\vec{\psi_{ls}} = L_{ls}\vec{i_s} \tag{2.22}$$

$$\vec{\psi_{lr}} = L_{lr}\vec{i_r} \tag{2.23}$$

where $\vec{\psi_m}$ is the mutual flux linkage, $\vec{\psi_{ls}}$ is the stator flux leakage, and $\vec{\psi_{ls}}$ is the rotor flux leakage, and L_m, L_{ls}, and L_{lr} are mutual, stator leakage, and rotor leakage inductances.

The induction machine circuit in space vector is expressed in Fig. 2.4.

2.2.1.1 Per Unit System
In per-unit system, the base voltage V_b and the base flux linkage ψ_b has a relationship $V_b = \omega_b \psi_b$. Per unitizing (2.11) and (2.18) leads to

$$\frac{\vec{v_s}}{V_b} = \frac{R_s \vec{i_s}}{Z_b I_b} + \frac{1}{\omega_b \psi_b}\frac{d\vec{\psi_s}}{dt} \tag{2.24}$$

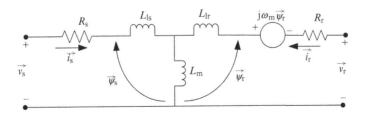

Figure 2.4 Induction machine circuit model in space vector.

$$\frac{\vec{v_r}}{V_b} = \frac{R_r \vec{i_r}}{Z_b I_b} + \frac{1}{\omega_b \psi_b}\frac{d\vec{\psi_r}}{dt} - j\frac{\omega_m}{\omega_b}\frac{\vec{\psi_r}}{\psi_b} \tag{2.25}$$

The per-unit model is as follows:

$$\vec{v_s} = R_s \vec{i_s} + \frac{1}{\omega_b}\frac{d\vec{\psi_s}}{dt} \tag{2.26}$$

$$\vec{v_r} = R_r \vec{i_r} + \frac{1}{\omega_b}\frac{d\vec{\psi_r}}{dt} - j\omega_m\vec{\psi_r} \tag{2.27}$$

$$\vec{\psi_s} = X_m(\vec{i_s} + \vec{i_r}) + X_{ls}\vec{i_s} \tag{2.28}$$

$$\vec{\psi_r} = X_m(\vec{i_s} + \vec{i_r}) + X_{lr}\vec{i_r} \tag{2.29}$$

where voltages, currents, flux linkages, rotating speed, resistances, and reactances are all in per unit. ω_b is the base angular frequency (377 rad/s).

Note that in per-unit system, we only deal with resistances and reactances. Values of these parameters are usually given based on rated machine power and voltages. There is no need to deal with inductances.

2.2.2 Induction Machine Modeling in Complex Vector

We now proceed to model an induction machine in a rotating reference frame (dq reference frame). The rotating speed is the flux rotating speed or the angular speed of the stator currents ω_s. Note the resulting model is the model after Park's transformation.

We already know that a set of three-phase-balanced sinusoidal currents or voltages can be converted to a space vector with constant magnitude and constant speed. Therefore, if we view the space vector from a rotating reference frame with the same speed, the space vector is now a stationary vector, or a complex vector. For a set of balanced stator currents or voltages, we have

$$\vec{V} = \overline{V}e^{j\omega_s t} \tag{2.30}$$

$$\overline{V} = \vec{V}e^{-j\omega_s t} \tag{2.31}$$

$\overline{V}$ is a complex vector. The complex vector can be expressed as $v_q - jv_d$, where q-axis leads d-axis by 90 degree.

The stator voltage and stator flux linkage relationship in space vector viewed from a stationary reference frame can now be transformed into a relationship expressed in complex vector viewed from the *dq* reference frame.

$$\overrightarrow{V_s}(t) = R_s\,\overrightarrow{i_s}\,(t) + \frac{\mathrm{d}}{\mathrm{d}t}\overrightarrow{\psi_s}(t)$$

$$\overline{V_s}e^{j\omega_s t} = R_s\overline{i_s}e^{j\omega_s t} + \frac{\mathrm{d}}{\mathrm{d}t}(\overline{\psi_s}e^{j\omega_s t})$$

$$= R_s\overline{i_s}e^{j\omega_s t} + (\frac{\mathrm{d}}{\mathrm{d}t}\overline{\psi_s})e^{j\omega_s t} + j\omega\overline{\psi_s}e^{j\omega_s t}$$

$$\Rightarrow \overline{V_s} = R_s\overline{i_s} + \frac{\mathrm{d}}{\mathrm{d}t}\overline{\psi_s} + j\omega_s\overline{\psi_s}$$

Similarly, for the rotor winding EMF and rotor flux linkage relationship, we have

$$\overrightarrow{V_r^r}(t) = R_r\,\overrightarrow{i_r^r}\,(t) + \frac{\mathrm{d}}{\mathrm{d}t}\overrightarrow{\psi_r^r}$$

$$\overline{V_r}e^{j\omega_r t} = R_r\overline{I_r}e^{j\omega_r t} + \frac{\mathrm{d}}{\mathrm{d}t}(\overline{\psi_r}e^{j\omega_r t})$$

where ω_r is the frequency of the rotor voltages and currents.

$$\Rightarrow \overline{V_r} = R_r\overline{I_r} + \frac{\mathrm{d}}{\mathrm{d}t}\overline{\psi_r} + j(\omega_s - \omega_m)\overline{\psi_r}$$

The relationship of the flux linkages and the currents are listed as follows:

$$\overline{\psi_s} = \overline{\psi_m} + \overline{\psi_{ls}} = L_m(\overline{I_s} + \overline{I_r}) + L_{ls}\overline{I_s}$$

$$\overline{\psi_r} = \overline{\psi_m} + \overline{\psi_{lr}} = L_m(\overline{I_s} + \overline{I_r}) + L_{lr}\overline{I_r}$$

The circuit model in the *dq* reference frame is shown in Fig. 2.5.

At steady state, the derivatives of the flux linkages are assumed to be zeros. Then we have

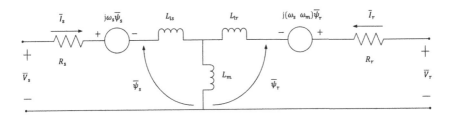

Figure 2.5 *Induction machine circuit model in the* dq *frame.*

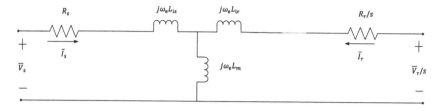

Figure 2.6 Induction motor steady-state equivalent circuit.

$$\overline{V}_s = R_s\overline{i}_s + j\omega_s\overline{\psi}_s$$

$$= (R_s + j\omega_s L_{ls})\overline{I}_s + j\omega_s L_m(\overline{I}_s + \overline{I}_r)$$

$$\overline{V}_r = R_r\overline{I}_r + js\omega_s(\overline{I}_r L_{lr} + (\overline{I}_s + \overline{I}_r)L_m)$$

$$= (R_r + js\omega_s L_{lr})\overline{I}_r + js\omega_s(\overline{I}_s + \overline{I}_r)L_m$$

The steady-state circuit model is illustrated in Fig. 2.6.

2.2.2.1 Model in Per-Unit System
The model in per-unit system can be expressed as follows:

$$\overline{V}_s = R_s\overline{i}_s + \frac{1}{\omega_b}\frac{d\overline{\psi}_s}{dt} + j\omega_s\overline{\psi}_s \tag{2.32}$$

$$\overline{V}_r = R_r\overline{I}_r + \frac{1}{\omega_b}\frac{d\overline{\psi}_r}{dt} + j(\omega_s - \omega_m)\overline{\psi}_r \tag{2.33}$$

$$\overline{\psi}_s = X_m(\overline{I}_s + \overline{I}_r) + X_{ls}\overline{I}_s = X_s\overline{I}_s + X_m\overline{I}_r \tag{2.34}$$

$$\overline{\psi}_r = X_m(\overline{I}_s + \overline{I}_r) + X_{lr}\overline{I}_r = X_m\overline{I}_s + X_r\overline{I}_r \tag{2.35}$$

where $X_s = X_m + X_{ls}$ and $X_r = X_m + X_{lr}$.

2.2.3 Swing Equation
According to Newton's second law for a rotating mass, the acceleration angular speed $\dot{\omega}_m$ is proportional to the net torque applied. The resulting relationship of the rotor speed ω_m, the mechanical torque T_m, and the electromagnetic torque T_e generated from the electromagnetic field is expressed as follows:

$$J\frac{d\omega_m}{dt} = T_e - T_m \tag{2.36}$$

where J is the inertia constant in N m s^2/rad. Torque is in standard unit N m, and ω_m is in radians/s.

It is desirable to have the equation in per-unit system. At both sides, the reciprocal of the base torque T_b will be multiplied. Note that

$$T_b = \frac{P_b^{3\phi}}{\omega_{mb}}$$

where ω_{mb} is the base mechanical speed and $P_b^{3\phi}$ is the base power for the three-phase system.

$$J\frac{d\omega_m}{dt} \cdot \frac{\omega_{mb}}{P_b^{3\phi}} = \frac{T_e}{T_b} - \frac{T_m}{T_b} \tag{2.37}$$

$$J\frac{\omega_{mb}^2}{P_b^{3\phi}} \cdot \frac{d\omega_{mpu}}{dt} = T_{epu} - T_{mpu} \tag{2.38}$$

Define $H \triangleq \frac{J\omega_{mb}^2}{2P_b^{3\phi}}$ and treat the variables as the per-unit system variables, we have

$$2H\frac{d\omega_m}{dt} = T_e - T_m \tag{2.39}$$

where H is in s pu, T_e and T_m are in pu.

To relate (2.39) with the electromagnetic equations, it is necessary to express the electromagnetic torque T_e in flux linkages and/or currents. Based on the physical mechanism, the torque is proportional to the magnitude of the cross product of two variables among the three (the mutual flux linkage, the stator current, and the rotor current). Therefore

$$T_e \propto \overrightarrow{\psi_m} \times \overrightarrow{i_s} \propto \overrightarrow{i_s} \times \overrightarrow{i_r} \tag{2.40}$$

The exact expression of electromagnetic torque can be found based on energy expression (Chapter 4, [1]). Here we will derive the expression based on power expression.

The power goes into the air gap can be expressed as

$$P_g = 3V_m I_s \sin\theta \tag{2.41}$$

where V_m and I_s are the RMS value of the per-phase air gap voltage and stator current, and θ is the angle between the stator current space vector and the air gap flux shown in Fig. 2.7.

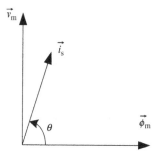

Figure 2.7 Space vectors of the stator current, air gap flux, and its induced EMF in the stator circuits.

Note that the magnitude of the space vectors or the complex vectors are the amplitude of the waveform. Therefore

$$P_g = \frac{3}{2}|\overrightarrow{V_m}||\overrightarrow{i_s}|\sin\theta \tag{2.42}$$

$$= \frac{3}{2}\omega_s|\overrightarrow{\psi_m}||\overrightarrow{i_s}|\sin\theta \tag{2.43}$$

Note that $T_e = \frac{P_m}{\omega_m} = \frac{P_g}{2\omega_s/P}$ where P is the pole numbers, therefore

$$T_e = \frac{3}{2}\frac{P}{2}|\overrightarrow{\psi_m}||\overrightarrow{i_s}|\sin\theta = \frac{3}{2}\frac{P}{2}\mathcal{R}(j\overrightarrow{\psi_m}\,\overrightarrow{i_s}^*) \tag{2.44}$$

$$= \frac{3}{2}\frac{P}{2}L_m|\overrightarrow{i_s}+\overrightarrow{i_r}||\overrightarrow{i_s}|\sin\theta \tag{2.45}$$

$$= \frac{3}{2}\frac{P}{2}L_m|\overrightarrow{i_s}||\overrightarrow{i_r}|\sin\theta_{sr} = \frac{3}{2}\frac{P}{2}L_m\mathcal{R}(j\overrightarrow{i_s}^*\overrightarrow{i_r})$$

$$= \frac{3}{2}\frac{P}{2}L_m\mathcal{R}(j\overline{I_s}^*\overline{I_r}) \tag{2.46}$$

$$= \frac{3}{2}\frac{P}{2}L_m(i_{qs}i_{dr}-i_{ds}i_{qr}) \tag{2.47}$$

$$= \frac{3}{2}\frac{P}{2}\frac{X_m}{\omega_s}(i_{qs}i_{dr}-i_{ds}i_{qr}) \tag{2.48}$$

In per-unit system, the torque expression becomes

$$T_e = X_m(i_{qs}i_{dr} - i_{ds}i_{qr}) \tag{2.49}$$

2.3 DFIG MODELING

In squirrel cage induction machines, the rotor circuits are short circuited. The type of machine used in DFIG wind turbine is wound-rotor induction machine, where the rotor circuits are not shorted. For DFIG, the rotor side will be connected to a back-to-back converter and tied back to the grid. For voltage-source converters (VSCs), the output voltage at the AC side consists of fundamental component and high-order harmonic components. The higher-order harmonic components can be easily filtered out by the rotor circuit inductors and have little influence on currents. The rotor currents in DFIG are almost sinusoidal. Therefore, in analytical models, only the fundamental components will be considered and the rotor voltage $\overrightarrow{v_r}$ is treated as a controllable voltage source with varying voltage magnitude and frequency.

The magnitude, frequency, and angle of the voltage source converters will be controlled through converter controls. This subject will be discussed in Chapter 3. In this chapter, beyond the electric machine model, the wind turbine aerodynamic model and the maximum power point tracking will be described as follows.

2.3.1 Wind Turbine Aerodynamics Model

The dynamic output mechanical torque of the wind turbine is expressed as [2]

$$T_m = \frac{1}{2}\rho A R C_p V_w^2 / \lambda \tag{2.50}$$

where ρ is the air density (kg/m^{-3}), A is the blade sweep area (m^2), R is the rotor radius of wind turbine (m), and V_w is the wind speed (m/s). C_p is the power coefficient of the blade which is a function of the blade pitch angle θ and the tip speed ratio as

$$C_p = \frac{1}{2}\left(\frac{RC_f}{\lambda} - 0.022\theta - 2\right)e^{-0.255\frac{RC_f}{\lambda}} \tag{2.51}$$

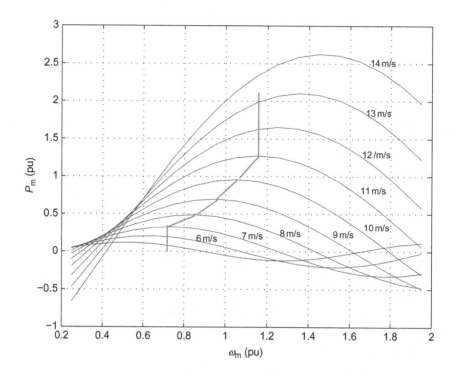

Figure 2.8 Mechanical power, rotor speed, and wind speed relationship.

where θ is the pitch angle and the tip speed ratio λ is

$$\lambda = \frac{\omega_m R}{V_w} \tag{2.52}$$

where ω_m is the mechanical angular velocity (rad/s).

The power, rotating speed and wind speed relationship when the pitch angle is zero for a 2 MW wind turbine with 200-ft blade radius is presented in Fig. 2.8. The wind speed that corresponding to the synchronous rotor speed is about 24 miles/h or 10.33 m/s.

We can come up with a lookup table (Table 2.1) giving the different wind speeds, the corresponding optimum rotor speed ω_m, the maximum mechanical power P_m, and the corresponding shaft toque T_m. The desired optimum rotor speed can be achieved by torque or power control through the rotor-side converter. This topic will be further explained in Chapter 3.

The continuous part of the maximum power point line can be expressed as $P_m^* = k\omega_m^3 = k(1 - s)^3$. This relationship will be useful to compute the

Table 2.1 Rotor Shaft Speed and Mechanical Power Lookup Table						
V_W m/s	7	8	9	10	11	12
ω_m	0.75	0.85	0.95	1.05	1.15	1.25
P_m	0.32	0.49	0.69	0.95	1.25	1.60
$T_m = \frac{P_m}{\omega_m}$	0.43	0.58	0.73	0.90	1.09	1.28

operating condition of a wind energy system. When steady-state calculation of a DFIG is investigated, this relationship is used in steady-state calculation (see Section 7.1).

2.4 EXAMPLES

Two examples are presented in this section to demonstrate the use of the analytical models. The first example illustrates the free acceleration of an induction machine. The machine is at standstill state with no load torque applied. Nominal voltage will be applied at its stator. The machine will accelerate to reach a nominal speed. This process of acceleration is call "free acceleration." The second example illustrates the consequence of DFIG stator voltage drop in rotor currents.

2.4.1 Example 1: Free Acceleration
Example 1. In Matlab/Simulink, build a simulation model to demonstrate the dynamics of free acceleration of an induction machine. Hint: please use complex vector-based model.

In numerical simulation, variables in real domain are used. A complex vector will be treated as two real variables. Throughout this book, the q-axis is leading the d-axis by 90°. This convention is the same as that in [3]. A complex vector can be expressed as qd-components as follows:

$$\bar{f} = f_q - jf_d$$

Using the model in complex vector and separating q-axis and d-axis components, the voltage equations of an induction machine can be written in terms of the currents as shown in (2.4.1). Note that the zero-sequence circuits are also added. Since zero-sequence currents do not introduce magnetic field, the zero-sequence circuits of stator and rotor are decoupled. Note that all variables are in per unit.

$$
\begin{bmatrix}
v_{qs} \\
v_{ds} \\
v_{0s} \\
v_{qr} \\
v_{dr} \\
v_{0r}
\end{bmatrix}
=
$$

$$
\begin{bmatrix}
R_s + \frac{p}{\omega_b}X_s & \frac{\omega_s}{\omega_b}X_s & 0 & \frac{p}{\omega_b}X_m & \frac{\omega_s}{\omega_b}X_m & 0 \\
-\frac{\omega_s}{\omega_b}X_s & R_s + \frac{p}{\omega_b}X_s & 0 & -\frac{\omega_s}{\omega_b}X_m & \frac{p}{\omega_b}X_m & 0 \\
0 & 0 & R_s + \frac{p}{\omega_b}X_{ls} & 0 & 0 & 0 \\
-\frac{p}{\omega_b}X_m & \frac{\omega_s-\omega_m}{\omega_b}X_m & 0 & R_r + \frac{p}{\omega_b}X_r & \frac{\omega_s-\omega_m}{\omega_b}X_r & 0 \\
-\frac{\omega_s-\omega_m}{\omega_b}X_m & \frac{p}{\omega_b}X_m & 0 & -\frac{\omega_s-\omega_m}{\omega_b}X_r & R_r + \frac{p}{\omega_b}X_r & 0 \\
0 & 0 & 0 & 0 & 0 & R_r + \frac{p}{\omega_b}X_{lr}
\end{bmatrix}
\begin{bmatrix}
i_{ds} \\
i_{qs} \\
i_{0s} \\
i_{qr} \\
i_{dr} \\
i_{0r}
\end{bmatrix}
\tag{2.53}
$$

The air gap flux linkages ψ_{qm} and ψ_{dm} can be expressed as

$$
\psi_{qm} = X_m(i_{qs} + i_{qr}) \tag{2.54}
$$
$$
\psi_{dm} = X_m(i_{ds} + i_{dr}) \tag{2.55}
$$

and the electromagnetic torque T_e can be expressed as

$$
T_e = \psi_{qm}i_{dr} - \psi_{dm}i_{qr} \tag{2.56}
$$

Assume that the reference frame is the synchronous reference frame and that all quantities are in per-unit value. Equation (2.53) can be further written in the form

$$
\dot{X} = AX + BU \tag{2.57}
$$

where $X = [i_{qs}, i_{ds}, i_{os}, i_{qr}, i_{dr}, i_{0r}]^T$, and

$$
B =
\begin{bmatrix}
\frac{X_s}{\omega_b} & 0 & 0 & \frac{X_m}{\omega_b} & 0 & 0 \\
0 & \frac{X_s}{\omega_b} & 0 & 0 & \frac{X_m}{\omega_b} & 0 \\
0 & 0 & \frac{X_{ls}}{\omega_b} & 0 & 0 & 0 \\
\frac{X_m}{\omega_b} & 0 & 0 & \frac{X_r}{\omega_b} & 0 & 0 \\
0 & \frac{X_m}{\omega_b} & 0 & 0 & \frac{X_r}{\omega_b} & 0 \\
0 & 0 & 0 & 0 & 0 & \frac{X'_{lr}}{\omega_b}
\end{bmatrix}^{-1}
\tag{2.58}
$$

$$A = -B \begin{bmatrix} R_s & \frac{\omega_s}{\omega_b}X_s & 0 & 0 & \frac{\omega_s}{\omega_b}X_m & 0 \\ -\frac{\omega_s}{\omega_b}X_s & R_s & 0 & -\frac{\omega_s}{\omega_b}X_m & 0 & 0 \\ 0 & 0 & R_s & 0 & 0 & 0 \\ 0 & \frac{\omega_s-\omega_m}{\omega_b}X_m & 0 & r'_r & \frac{\omega_s-\omega_m}{\omega_b}X_r & 0 \\ -\frac{\omega_s-\omega_m}{\omega_b}X_m & 0 & 0 & \frac{\omega_s-\omega_m}{\omega_b}X_r & R_r & 0 \\ 0 & 0 & 0 & 0 & 0 & R_r \end{bmatrix}$$

$$(2.59)$$

The swing equation is

$$T_e = 2H\dot{\omega}_m + T_m \qquad (2.60)$$

where T_m is the mechanical torque and H is the inertia.

In summary, differential equations (2.57) and (2.60) represent the induction machine with its seventh-order model.

2.4.1.1 Induction Machine Simulink Model

The Simulink model in terms of the state space equations (2.57) is shown in Fig. 2.10. In this model block, the inputs are the voltage vector and rotor speed and the output is a current vector. This model is quite simple and easy to understand. It saves not only on model building time but also debugging time. The rotor speed dynamics (2.60) is modeled by another integrator block, shown in Fig. 2.11.

Note that to obtain a flat run, the initial state variables should be given. The known initial condition should be passed to the dynamic simulation blocks through the integrator unit. Figure 2.9 shows an integrator block in Matlab/Simulink. Generally speaking, any integrator should have the initial values setup. This specific integrator corresponds to the one of current integrator. Therefore, I_{0ind} should be computed through steady-state calculation and then passed to the dynamic simulation blocks.

The rotor speed will be fed back to the input of the block in Fig. 2.10. The induction machine serves as a current source to the network and the output from the network is the voltage vector. Thus, the induction machine and the power system network are interconnected, and as long as the initial condition is set, dynamic simulation can be performed.

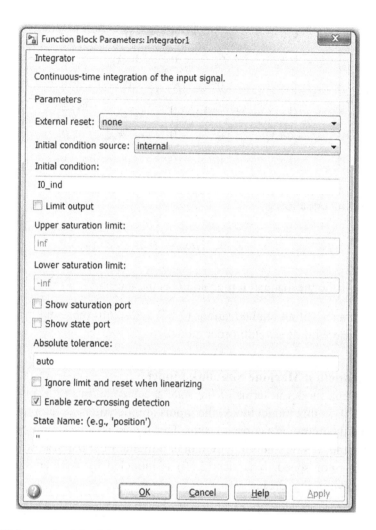

Figure 2.9 State equations in Simulink.

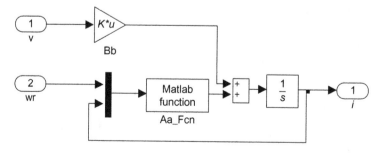

Figure 2.10 Swing equation in Simulink.

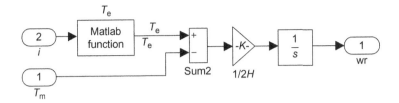

Figure 2.11 Initialization through integrators.

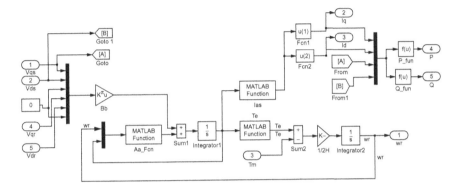

Figure 2.12 Induction generator modeling in Simulink.

The entire induction generator model block consists both the swing equation and the current state space model. The detailed model in Simulink is shown in Fig. 2.12.

The developed Matlab/Simink model is benchmarked by comparing the free acceleration of an induction machine from Krause' book [3]. To simulate the free acceleration, the stator voltages v_{qs} is set to 1 pu and all other voltages v_{ds}, v_{qr}, and v_{dr} are set to zero. The rotor is short circuited. The simulation results shown in Fig. 2.13 are same as those from the textbook.

2.4.2 Example 2: DFIG Stator Voltage Drop
The second example on the consequences of DFIG stator voltage drop is first presented in [4].

Example 2. Explain the effect of a three-phase voltage drop (to zero) in the stator voltage of a DFIG in rotor voltage.

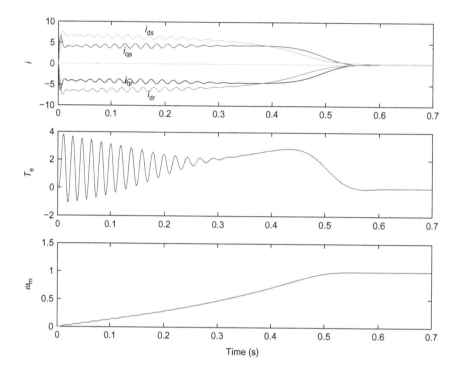

Figure 2.13 Free acceleration characteristics of a 10 hp induction motor in the synchronously rotating reference frame.

Hint.

- use space vector model;
- find the relationship between the stator voltage and the stator flux assuming the rotor current is zero. From this relationship, a full dip in stator voltage is equivalent to a decay in stator flux;
- express rotor voltage in terms of stator flux and rotor current;
- ignore the effect of rotor current and investigate the effect of stator flux decay on rotor voltage.

Solution. First we examine the stator voltage and stator flux relationship.

$$\vec{v_s} = R_s \vec{i_s} + \frac{d\vec{\psi_s}}{dt} \tag{2.61}$$

$$\vec{\psi_s} = L_m(\vec{i_s} + \vec{i_r}) + L_{ls} \vec{i_s} \tag{2.62}$$

With the assumption of rotor current is zero, the stator flux is only related to the stator current.

$$\vec{\psi_s} \approx L_s \vec{i_s} \tag{2.63}$$

Therefore, we have

$$\vec{v_s} \approx \frac{R_s}{L_s}\vec{\psi_s} + \frac{d\vec{\psi_s}}{dt} \tag{2.64}$$

Before the drop occurs (t_0^-), if the effect of the stator resistance can be ignored, the stator flux can be found based on the stator voltage.

$$\vec{v_s}(t_0^-) = |V_s|e^{j\omega_s t_0} \tag{2.65}$$

$$\vec{\psi_s}(t_0^-) = -j\frac{|V_s|}{\omega_s}e^{j\omega_s t_0} \tag{2.66}$$

After the drop occurs (t_0^+), the stator voltage will have an immediate change. However, the flux will not change abruptly. Therefore,

$$\vec{v_s}(t_0^+) = 0 \tag{2.67}$$

$$\vec{\psi_s}(t_0^+) = -j\frac{|V_s|}{\omega_s}e^{j\omega_s t_0} \tag{2.68}$$

Based on the dominant dynamic equation (2.64), when stator voltage is zero, the stator flux will decay as time evolves.

$$\vec{\psi_s}(t) = \vec{\psi_s}(t_0)e^{-\frac{R_s}{L_s}(t-t_0)} = -j\frac{|V_s|}{\omega_s}e^{j\omega_s t_0}e^{-\frac{R_s}{L_s}(t-t_0)}, \quad \text{for } t > t_0^+ \tag{2.69}$$

We now proceed to find the relationship between the rotor voltage $\vec{v_r}$ and the stator flux $\vec{\psi_s}$. Note that the rotor voltage is related the rotor flux through Faraday's law. Hence we would like to first express the rotor flux by the stator flux, and then to find the relationship between the rotor voltage and the stator flux.

Since

$$\vec{\psi_s} = L_m \vec{i_r} + L_s \vec{i_s} \tag{2.70}$$

$$\vec{\psi_r} = L_m \vec{i_s} + L_r \vec{i_r}, \tag{2.71}$$

further the rotor current can be ignored; therefore, we have

$$\vec{\psi_r} = \frac{L_m}{L_s}\vec{\psi_s}.$$ (2.72)

Replacing the rotor flux by the stator flux in the rotor voltage expression and ignoring the effect of the rotor current, we have

$$\vec{v_r} \approx -j\omega_m\vec{\psi_r} + \frac{d\vec{\psi_r}}{dt} = \frac{L_m}{L_s}\left(j\omega_m\vec{\psi_s} + \frac{d\vec{\psi_s}}{dt}\right)$$ (2.73)

Substituting $\vec{\psi_s}$ by (2.69), we have

$$\vec{v_r}(t) = \frac{L_m}{L_s}\vec{\psi_s}(t_0)e^{-\frac{R_s}{L_s}(t-t_0)}\left(-\frac{R_s}{L_s} - j\omega_m\right)$$ (2.74)

At $t = t_0^+$, then

$$\vec{v_r}(t_0^+) = \frac{L_m}{L_s}\vec{\psi_s}(t_0)\left(-\frac{R_s}{L_s} - j\omega_m\right)$$ (2.75)

$$|\vec{v_r}(t_0^+)| \approx \frac{|V_s|}{\omega_s}\frac{L_m}{L_s}\omega_m \approx (1-s)|V_s|$$ (2.76)

where s is the slip and $s = \frac{\omega_m - \omega_s}{\omega_s}$.

Remarks. This phenomenon can be explained as follows: due to a sudden voltage drop, the previous rotating flux is no longer rotating. This flux seen by the rotor has a slip speed of $s\omega_s$ previously and now has a speed of $-\omega_m$. Therefore, previously, the induced rotor voltage was $s\omega_s|\psi_m| \approx s|V_s|$, and now it becomes $\omega_m|\psi_m| \approx (1-s)|V_s|$. The rotor voltage will have a sudden increase.

A simulation case study is conducted to examine the effect of full dip stator voltage. A demo DFIG-based wind farm system in Matlab/SimPowerSystems toolbox is used. The DFIG model is an average model with the converters modeled as controllable voltage sources while the power electronic switching details are ignored. The system is modified to match the assumption made in Example 2 that the rotor current can be

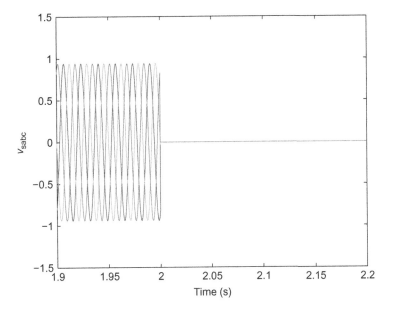

Figure 2.14 Three-phase stator voltages.

ignored. A three-phase large resistor is added from the rotor side converter to the DFIG's rotor input point. That way, the rotor current will be greatly reduced and the effect of the rotor current can be ignored. A three-phase breaker connects the stator bus to the ground. The initial status of the breaker is open. At 2 s, the breaker closes and the stator voltage has a full dip and goes to zero afterwards. Figures 2.14–2.16 demonstrate the transients in stator voltages, rotor voltages, and stator currents. The variables are all in per units.

Figure 2.14 shows the stator voltages. At $t = 2$ s, the stator voltages drop to zero. Figure 2.15 shows the rotor voltages. At $t = 2$ s, the rotor voltages rise sharply. The DFIG is operating at a supersynchronous speed (slip <0). Therefore, the rotor voltage magnitude exceeds the stator voltage magnitude at the beginning of the transient. The rotor voltages then decay to zero based on (2.74). Figure 2.16 shows the stator currents. The stator current space vector is proportionally related to the stator flux linkage space vector due to the omission of the rotor current. Therefore, the stator currents decay to zero just as the stator flux linkages decay to zero.

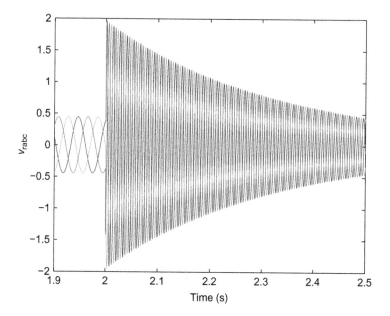

Figure 2.15 Three-phase rotor voltages.

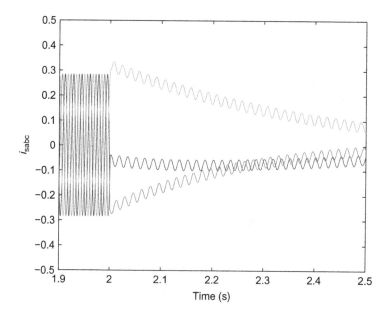

Figure 2.16 Three-phase stator currents.

REFERENCES

[1] E. Fitzgerald, C. Kingsley, S. Umans, Electric Machinery, McGraw Hill, New York, 1990.

[2] Y. Lei, A. Mullane, G. Lightbody, R. Yacamini, Modeling of the wind turbine with a doubly fed induction generator for grid integration studies, IEEE Trans. Energy Convers. 21 (2006) 257-264.

[3] P.C. Krause, O. Wasynczuk, S.D. Sudhoff, S. Pekarek, Analysis of Electric Machinery and Drive Systems, vol. 75, John Wiley & Sons, New York, 2013.

[4] J. Lopez, P. Sanchis, X. Roboam, L. Marroyo, Dynamic behavior of the doubly fed induction generator during three-phase voltage dips, IEEE Trans. Energy Convers. 22(3) (2007) 709-717.

Modeling of Doubly Fed Induction Generation (DFIG) Converter Controls

3.1 ROTOR FLUX-ORIENTED INDUCTION MACHINE CONTROL 35

3.1.1 Torque/Flux Control 36

3.1.2 Inner Current Control 38

3.2 DFIG ROTOR SIDE CONVERTER CONTROL 40

3.2.1 Outer Control 41

3.2.2 Inner Current Control 42

3.2.3 Maximum Power Point Tracking 44

3.3 GSC CONTROL 46

3.3.1 Outer Control 46

3.3.2 Inner Current Control 47

3.4 COMPLETE DFIG MODELING BLOCKS 48

3.5 EXAMPLES 49

3.5.1 Example 1: PSCAD Simulation of a Two-Level VSC
with Sine PWM 49

3.5.2 Example 2: DFIG Simulation 52

REFERENCES 54

In this chapter, the major effort is devoted to converter controls of DFIG. Differences between drive control and grid integration control are discussed before any presentation of controls. To refresh our memory on drive control, rotor flux-oriented control for an induction machine is first presented, followed by DFIG's rotor-side converter (RSC) control and grid-side converter (GSC) control. With a converter being considered as a controllable AC voltage source with a controllable frequency, magnitude, and phase angle, the entire DFIG system model (with converter controls) can now be built in the *dq*-reference frame.

Modeling and Analysis of Doubly Fed Induction Generator Wind Energy Systems.
http://dx.doi.org/10.1016/B978-0-12-802969-5.00003-2

This chapter presents two simulation examples. The first one is a simulation example in PSCAD with power electronic switching details. The purpose of the demonstration is to show that converter voltage outputs after filters are sinusoidal. Readers can then build visual connections between control of *abc* sinusoidal waveforms versus vector control or control of *dq* variables. The second one is a demonstration of DFIG converter control using the overall DFIG dynamic model.

3.1 ROTOR FLUX-ORIENTED INDUCTION MACHINE CONTROL

Before discussing DFIG converter control, let us review a few concepts related to an induction machine control. Interested readers can refer AC drive textbooks such as [1], for a variety of drive controls, including rotor flux-oriented control and stator flux-oriented control.

From a control engineer's perspective, the first thing to notice is the input/output of the plant: what will be adjusted and which measurements will be used as signals. In the case of induction machine control, we would like to adjust the stator voltage to realize speed control. Therefore, stator voltage is the input of the plant while speed is the output. The stator voltage can be generated by a DC/AC voltage source converter. Through adjusting pulse width modulation (PMW)'s control signals, the output voltage can have a controllable magnitude, frequency, and angle. Note that the output voltage directly from the converter is of discrete voltage levels. It can be decomposed into a fundamental frequency component and higher-order harmonic components with the frequencies at the order of the switching frequency. For an IGBT voltage source converter, the switching frequency is more than ten times of the fundamental frequency. For example, for a 50-Hz system, the switching frequency could be 750 Hz.

Averaging is a modeling and analysis technique popularly applied in power electronics [2]. With switches, converters generate voltages with discrete levels. Through averaging the converter output voltage over a switching period, the higher-order harmonics will all be gone. Only the lower-order harmonics will be considered in modeling. Averaging technique greatly simplifies modeling. It is also reasonable to ignore higher-order harmonics since an inductive filter after the converter or the inductance in windings can get rid of those harmonics. Another benefit of average models is that discrete switching is now ignored from modeling. Average models are continuous dynamic models.

For the analytical models built throughout the textbook, the underlying assumption is that the models are average models. Ultimately, the converter control is expected to generate the stator voltage reference signals. These sinusoidal signals will be fed into the PWM of a VSC to generate pulses to six gates for a typical two-level voltage source converter.

Details of PWM and IGBT switching will not be discussed in this book as these can be found in a typical power electronics textbook. Converters are sensitive to over currents. Therefore, currents through a converter are expected to be regulated very quickly and follow their references. For an induction machine, a constant flux is desired. In summary, an induction machine control should include speed (torque) control, flux control, and current control. In addition, the current control should be much faster than the torque and flux control. Therefore, when designing current control, we can treat the flux and torque as constant, while when we design the torque or flux control, we can assume that the currents can follow their commands immediately. The current control design and the torque/flux control design can be done separately with these assumptions.

3.1.1 Torque/Flux Control

The principal of the decoupled torque/flux control is explained in the following paragraphs. From Chapter 2, we know that the torque can be expressed in terms of air gap flux linkage and the stator current.

$$T_e = \frac{3}{2}\frac{P}{2}\mathcal{R}\left(j\overline{\psi_m I_s}^*\right) \tag{3.1}$$

Note that the air gap flux linkage and the rotor flux linkage can be related as follows:

$$\overline{\psi_r} = L_r\overline{I_r} + L_m\overline{I_s} \tag{3.2}$$

$$= L_r(\overline{I_r} + \overline{I_s}) - L_{lr}\overline{I_s} \tag{3.3}$$

$$= \frac{L_r}{L_m}\overline{\psi_m} - L_{lr}\overline{I_s} \tag{3.4}$$

Therefore, the torque can be expressed in terms of the rotor flux linkage and the stator current.

$$T_e = \frac{3}{2}\frac{P}{2}\mathcal{R}\left(j\frac{L_m}{L_r}\left(\overline{\psi_r} + L_{lr}\overline{I_s}\right)\overline{I_s}^*\right) \tag{3.5}$$

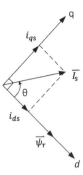

Figure 3.1 Current complex vector decomposed based on the rotor flux-oriented reference frame.

$$= \frac{3}{2}\frac{P}{2}\frac{L_{\mathrm{m}}}{L_r}\mathcal{R}\left(\overline{j\psi_r I_{\mathrm{s}}^*}\right) \tag{3.6}$$

$$= \frac{3}{2}\frac{P}{2}\frac{L_{\mathrm{m}}}{L_r}\left(\psi_{dr}i_{qs} - \psi_{qr}i_{ds}\right) \tag{3.7}$$

It is easily seen in Fig. 3.1 that when the d-axis is aligned with the rotor flux $\overrightarrow{\psi}_r$ ($\overline{\psi}_r$ can be decoupled as $\psi_{dr} = \hat{\psi}_r$ and $\psi_{qr} = 0$), $\overline{I}_{\mathrm{s}}$ can be decoupled as i_{qs} and i_{ds} and the electromagnetic torque T_{e} is only related to the q-axis stator current i_{qs}.

$$T_{\mathrm{e}} = k\hat{\psi}_r i_{qs} \tag{3.8}$$

where $k = \frac{3}{2}\frac{P}{2}\frac{L_{\mathrm{m}}}{L_r}$. If the rotor flux magnitude $\hat{\psi}_r$ is constant, then T_{e} and i_{qs} has a linear relationship. If we want to adjust the torque, we can adjust i_{qs} only. On the other hand, we also need to make sure that adjusting i_{qs} will not change the rotor flux magnitude. This can be proved by the following derivation.

The relationship between $\overline{\psi}_r$ and $\overline{I}_{\mathrm{s}}$ is shown below:

$$\overline{\psi}_r = L_r\overline{i}_r + L_{\mathrm{m}}\overline{i}_{\mathrm{s}} \Rightarrow \overline{i}_r = \frac{1}{L_r}(\overline{\psi}_r - L_{\mathrm{m}}\overline{i}_{\mathrm{s}}) \tag{3.9}$$

The relationship between the rotor voltage and the rotor flux is shown as follows:

$$\overline{V}_r = 0 = R_r\overline{i}_r + \frac{d\overline{\psi}_r}{dt} + j\omega_{sl}\overline{\psi}_r \tag{3.10}$$

where ω_{sl} is the slip frequency.

Therefore, for qd-axis variables, we find the following relationship:

$$\begin{cases} 0 = \frac{R_\mathrm{r}}{L_\mathrm{r}}(\psi_{qr} - L_\mathrm{m}i_{qs}) + \frac{\mathrm{d}\psi_{qr}}{\mathrm{d}t} + \omega_{sl}\psi_{dr} \\ 0 = \frac{R_\mathrm{r}}{L_\mathrm{r}}(\psi_{dr} - L_\mathrm{m}i_{ds}) + \frac{\mathrm{d}\psi_{dr}}{\mathrm{d}t} - \omega_{sl}\psi_{qr} \end{cases} \quad (3.11)$$

Since $\psi_{qr} = 0$, its derivative $\frac{\mathrm{d}\psi_{qr}}{\mathrm{d}t}$ also equals zero. In addition, at steady state, the flux is kept constant. Therefore $\frac{\mathrm{d}\psi_{dr}}{\mathrm{d}t} = 0$. Thus, we have

$$\omega_{sl} = \frac{R \cdot L_\mathrm{m}}{L_\mathrm{r} \cdot \hat{\psi}_\mathrm{r}} i_{qs} \quad (3.12)$$

$$\psi_{dr} = L_\mathrm{m}i_{ds} \quad (3.13)$$

The rotor flux is related to the d-axis stator current only. Therefore, changing i_{qs} will not impact the steady-state rotor flux magnitude. The aforementioned analysis demonstrates that the torque and flux can be controlled in a decoupled fashion. Given a constant rotor flux, torque is proportional to the q-axis stator current. To track a torque or flux reference, proportional integral controllers are employed to generate the qd-axis current commands. These commands will be tracked by the current controllers.

3.1.2 Inner Current Control

To derive a current control strategy, first of all, we need to find the plant model where the outputs of the plant are current signals and the inputs of the plant model are converter voltages. The control will be based on the rotor flux-oriented reference frame. Therefore, the plant model is also based on the rotor flux-oriented reference frame.

Let us review the relationship between flux linkages and currents.

$$\psi_{qr} = L_\mathrm{r}i_{qr} + L_\mathrm{m}i_{qs} \quad (3.14)$$

$$\psi_{dr} = L_\mathrm{r}i_{dr} + L_\mathrm{m}i_{ds} \quad (3.15)$$

Based on the condition of the rotor flux-oriented reference frame, we know that $\psi_{dr} = \hat{\psi}_r$ and $\psi_{qr} = 0$. Therefore, we have the following relationship between the stator current and the rotor current.

$$i_{qr} = -\frac{L_\mathrm{m}}{L_\mathrm{r}}i_{qs} \quad (3.16)$$

$$i_{dr} = -\frac{L_\mathrm{m}}{L_\mathrm{r}}\left(\hat{\psi}_\mathrm{r} - L_\mathrm{m}i_{ds}\right) \quad (3.17)$$

The stator flux linkages in the rotor flux-oriented reference frame can now be expressed by the stator currents and the rotor flux magnitude solely.

$$\psi_{qs} = \left(L_s - \frac{L_m^2}{L_r} \right) i_{qs} = \sigma L_s i_{qs} \tag{3.18}$$

$$\psi_{ds} = \left(L_s - \frac{L_m^2}{L_r} \right) i_{ds} + \frac{L_m}{L_r} \hat{\psi}_r = \sigma L_s i_{ds} + \frac{L_m}{L_r} \hat{\psi}_r \tag{3.19}$$

where $\sigma = 1 - \frac{L_m^2}{L_s L_r}$.

In the rotor flux-oriented reference frame, the relationship of the stator voltage and the stator current can then be found as follows:

$$v_{qs} = r i_{qs} + \sigma L_s \frac{d i_{qs}}{dt} + \sigma \omega L_s i_{ds} + \omega \frac{L_m}{L_r} \hat{\psi}_r \tag{3.20}$$

$$v_{ds} = r i_{ds} + \sigma L_s \frac{d i_{ds}}{dt} - \sigma \omega L_s i_{qs} \tag{3.21}$$

where ω is the rotor flux rotating speed. This speed should be the same as the electric frequency in the stator voltage and currents.

Feedforward techniques can be applied to design the current controllers for the qd-axis respectively. For simplicity, if we are not keen to have decoupled effect of qd-axis current tracking, we can also live with a feedback control without feedforward terms.

Define two variables

$$u_{qs} = v_{qs} - \sigma \omega L_s i_{ds} - \omega \frac{L_m}{L_r} \hat{\psi}_r \tag{3.22}$$

$$u_{ds} = v_{ds} + \sigma \omega L_s i_{qs}. \tag{3.23}$$

Then, we can have two plant models:

$$\frac{i_{qs}}{u_{qs}} = \frac{1}{r + \sigma L_s s} \tag{3.24}$$

$$\frac{i_{ds}}{u_{ds}} = \frac{1}{r + \sigma L_s s} \tag{3.25}$$

We can design the feedback controller in the format of PI control to achieve desired bandwidth. Assume that the PI controller is $K_P + K_i/s$. Then the open-loop transfer function (or loop gain) is

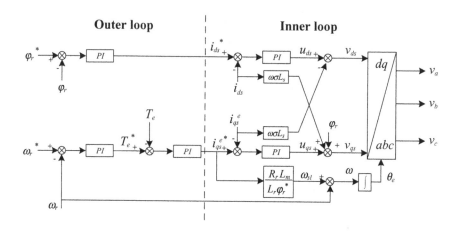

Figure 3.2 Induction machine control block diagram.

$$L(s) = \frac{K_P(s + K_P/K_i)}{s} \cdot \frac{1}{r + \sigma L_s s}.$$

Let $K_i/K_P = R/(\sigma L_s)$ so that the zero can cancel one of the poles. The loop gain now becomes:

$$L(s) = \frac{K_P}{\sigma L_s s}.$$

The closed-loop transfer function is

$$L_{cl}(s) = \frac{1}{1 + \frac{\sigma L_s}{K_p}s}$$

If the desired bandwidth is ω_B, then $\omega_B = \frac{K_P}{\sigma L_s}$. Based on this relationship, we can determine K_P and further determine K_i.

The overall control block diagram shown in Fig. 3.2 includes feedforward terms. The controller parameters can be obtained by trial and error or by analysis.

3.2 DFIG ROTOR SIDE CONVERTER CONTROL

The difference between a grid-integrated DFIG wind turbine and an induction machine resides in control objectives. For a DFIG, the converter controls are to regulate real power (or torque) and reactive power (or voltage) send to the grid through RSC and GSC's output voltages. RSC's and GSC's AC-side output voltages are the inputs to the plants and the outputs from the converter controls.

DFIG converter controls have been well documented in the literature, e.g., [3]. In this chapter, we will give a brief explanation of the vector control philosophy, the related plant models, and control design.

3.2.1 Outer Control

A RSC is connected to the rotor circuit. Therefore, the rotor currents should also be regulated to avoid overcurrent in the RSC. In that sense, we can reason that the inner current control for a RSC should be the rotor current control, while the outer control should be the real power (torque) and reactive power (ac voltage) control.

Unlike rotor flux-oriented control of an induction machine, DFIG's control relies on stator flux-oriented reference frame ($\psi_{qs} = 0$ or $v_{ds} = 0$). Since DFIG wind turbines are integrated to the grid and the grid voltage can be assumed as constant, the stator flux of the DFIG can be assumed as constant. In the literature, we may find other types of reference frames, e.g., grid flux (flux corresponding to the grid voltage) oriented reference frame, which is also reasonable.

We will express the torque by the stator flux and the rotor current to show that decoupling control is possible for a DFIG. Employing the same technique shown in the previous section, we find that

$$\overline{\psi}_s = L_s \overline{I}_s + L_m \overline{I}_r \tag{3.26}$$

$$= \frac{L_s}{L_m} \overline{\psi}_m - L_{ls} \overline{I}_r \tag{3.27}$$

Therefore, the torque expression can be found:

$$T_e = \frac{3}{2} \frac{P}{2} \frac{L_m}{L_s} \left(\psi_{qs} i_{dr} - \psi_{ds} i_{qr} \right) \tag{3.28}$$

$$= -\frac{3}{2} \frac{P}{2} \frac{L_m}{L_s} \psi_{ds} i_{qr} \tag{3.29}$$

Note that in this book, we adopt the motor convention. Therefore, when we consider this is a generator, the output power is related to $-T_e$. The output real power P_s from the stator circuit while ignoring the copper loss can be expressed as

$$P_s = -\omega_e T_e = \frac{3}{2}\frac{P}{2}\omega_e \frac{L_m}{L_s}\psi_{ds}i_{qr} \tag{3.30}$$

where ω_e is the synchronous mechanical speed and $\omega_s = \frac{P}{2}\omega_e$ where ω_s is the electric frequency of the stator.

The reactive power from the stator circuit can be expressed as

$$Q_s = -\frac{3}{2}(v_{qs}i_{ds} - v_{ds}i_{qs}) \approx -\frac{3}{2}\frac{P}{2}\omega_e\psi_{ds}i_{ds} \tag{3.31}$$

Note that $\psi_{ds} = L_m i_{dr} + L_s i_{ds}$. When the stator flux is constant, increase (decrease) in i_{dr} results in increase (decrease) in $-i_{ds}$ or Q_s.

Remarks. The output real and reactive power from the stator circuit can be controlled via i_{qr} and i_{dr} respectively. The outer control can be torque/reactive power or torque/AC voltage. In that case, the generator's torque is also only related to i_{qr}.

3.2.2 Inner Current Control

From the outer power control, rotor current references (of the stator flux reference frame) will be generated. It is through the inner current control that the rotor current commands will be followed by the rotor currents. Feedback control will be employed to realize the command tracking.

We will start from the rotor voltage and rotor flux linkage relationship, then derive the rotor voltage and rotor current relationship, and finally find the plant model for current control.

The complex vector model of the rotor voltage/rotor flux linkage is expressed as follows:

$$\overline{V}_r = r_r \overline{I}_r + \frac{d\overline{\psi}_r}{dt} + j\omega_{sl}\overline{\psi}_r \tag{3.32}$$

$$\begin{cases} v_{qr} = r_r i_{qr} + \frac{d\psi_{qr}}{dt} + \omega_{sl}\psi_{dr} \\ v_{dr} = r_r i_{dr} + \frac{d\psi_{dr}}{dt} - \omega_{sl}\psi_{qr} \end{cases} \tag{3.33}$$

Use the condition of stator flux orientation, we have

$$0 = \psi_{qs} = L_s i_{qs} + L_m i_{qr} \tag{3.34}$$

$$\hat{\psi}_s = \psi_{ds} = L_s i_{ds} + L_m i_{dr}. \tag{3.35}$$

Replace the stator currents in the rotor flux linkage expressions by the rotor currents and stator flux linkages:

$$\psi_{qr} = L_r i_{qr} + L_m i_{qs} = \left(L_r - \frac{L_m^2}{L_s} \right) i_{qr} = \sigma L_r i_{qr} \tag{3.36}$$

$$\psi_{dr} = L_r i_{dr} + L_m i_{ds} = L_r i_{dr} + L_m \frac{\psi_{ds} - L_m i_{dr}}{L_s} = \sigma L_r i_{dr} + \frac{L_m}{L_s} \hat{\psi}_s \tag{3.37}$$

The rotor voltages can now be expressed by the rotor currents only:

$$v_{qr} = r_r i_{qr} + \sigma L_r \frac{d i_{qr}}{dt} + \omega_{sl} \left(\sigma L_r i_{dr} + \frac{L_m}{L_s} \hat{\psi}_s \right) \tag{3.38}$$

$$v_{dr} = r_r i_{dr} + \sigma L_r \frac{d i_{dr}}{dt} - \omega_{sl} \sigma L_r i_{qr} \tag{3.39}$$

Introduce two virtual variables

$$u_{qr} = v_{qr} - \omega_{sl} \left(\sigma L_r i_{dr} + \frac{L_m}{L_s} \hat{\psi}_s \right) \tag{3.40}$$

$$u_{dr} = v_{dr} + \omega_{sl} \sigma L_r i_{qr} \tag{3.41}$$

The plant models can be found as:

$$i_{qr} = \frac{1}{r_r + \sigma L_r s} u_{qr} \tag{3.42}$$

$$i_{dr} = \frac{1}{r_r + \sigma L_r s} u_{dr} \tag{3.43}$$

Feedback controllers can be designed based on the above two first-order plant models to have desired bandwidths. After the feedback controllers, feedforward compensation should be added back to generate the desired rotor voltages.

The overall control diagram is presented in Fig. 3.3. The inner current control design and the output power control design are carried out in two separate steps with the underlying assumption: the dynamics of the current control is much faster than the power control. Separate control design will be carried out for inner current control and outer power control.

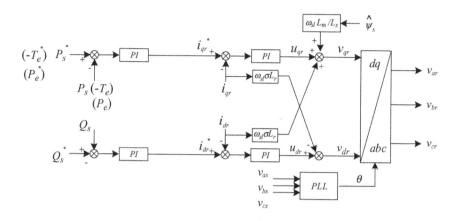

Figure 3.3 DFIG RSC control block diagram.

Indeed, we may be more interested in control of the entire export power or electromagnetic torque from DFIG instead of just the stator power. This can also be done by adjusting i_{qr} only. Recall that

$$P_s = -\omega_e T_e = \frac{3}{2}\frac{P}{2}\omega_e \frac{L_m}{L_s}\psi_{ds}i_{qr}. \tag{3.44}$$

$$P_e = -\omega_m T_e = (1+s)P_s \tag{3.45}$$

If the rotor speed varies much slower than the power control, then we can assume that the slip s is constant and to regulate the entire power P_e, we just need to adjust i_{qr}. For a short period of seconds, the wind speed can be assumed as constant. The rotor speed can also be considered as constant.

Additional notes: For the inner current control diagram, the feedforward compensation can be ignored. In that case, to follow a q-axis current command, both qd-axis voltages will be adjusted. Ignoring the feedback compensation makes control simpler and easy to implement. The advantages of feedforward compensation can be found in [4, Chapter 3], including faster start-up transient, decoupling with the AC system, and better disturbance rejection capability.

3.2.3 Maximum Power Point Tracking
Maximum power point tracking (MPPT) can be realized in the RSC control by adjusting the power or torque command. Suppose that the total power

from the DFIG will be controlled. The command of the total power will be generated through the MPPT control block. The input of the control block is the rotor speed ω_m. Through the lookup table, the optimum power corresponding to this speed will be generated. This power will be passed to the outer power control block as the power command.

The ability to get maximum power by the MPPT block is explained as follows. Figure 3.4 presents the wind speed, rotating speed, and mechanical power relationship. The red line (dark grey in print version) is the maximum power curve. Suppose that the wind speed is 9 m/s, the wind generator has a rotating speed lower than the optimum speed. The operating point is notated as Point A, where the optimum operating point is notated as Point B. According to the MPPT lookup table, the generated power command (Point A') will be less than the current mechanical power. Assuming that the power control dynamics are very fast, the wind turbine now endures a power unbalance: the mechanical power is greater than the electric power. The rotor will speed up until Point B is reached. At Point B, the mechanical power and the electric power command match each other. Similarly, when the wind turbine operates at a rotating speed greater than the optimum speed,

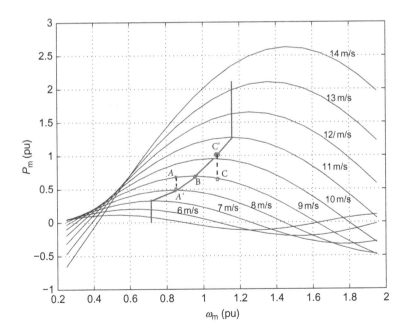

Figure 3.4 MPPT control explanation.

Point C, according to MPPT look up table, the power command should be the same at Point C'. The electric power is now greater than the mechanical power. The rotor should slow down until the operating point reaches B.

3.3 GSC CONTROL

The GSC is connected to the grid through a filter and/or a transformer. The GSC is expected to regulate the AC side voltage/reactive power and to keep the DC-link capacitor voltage constant. With a constant DC-link voltage, the power through the RSC will be the same as that through the GSC. Therefore, this control objective realizes power balance of the converters.

3.3.1 Outer Control

Let the q-axis of the reference frame align with the coupling point voltage $\overrightarrow{e}$ ($e_d = 0$) and notate the converter output voltage as $\overrightarrow{v}_g$ or $\overline{V}_g$. We can then express the real power and reactive power from the GSC to the coupling point as

$$P_g + jQ_g = \frac{3}{2}\overline{E}\overline{I}_g^*$$ (3.46)

$$P_g = \frac{3}{2}(e_q i_{qg} + e_d i_{dg}) = \frac{3}{2}e_q i_{qg}$$ (3.47)

$$Q_g = \frac{3}{2}(e_q i_{dg} - e_d i_{qg}) = \frac{3}{2}e_q i_{dg}$$ (3.48)

Therefore, if the coupling point voltage is kept constant (this should be the case for a grid-connected DFIG), real power and reactive power are linearly related to the q-axis and d-axis currents, respectively. We can again design decoupled real power and reactive power control. Keep in mind that the GSC control and RSC control should be coordinated. RSC control has the objective to track the stator active/reactive power commands. Then GSC control should take care of the DC-link voltage.

The DC-link capacitor voltage can be expressed in terms of the power from the RSC and the power leaving the GSC to the grid. The convention of P_r follows the rotor current convention, where injection to the rotor circuit is positive. The convention of P_g follows the GSC current convention, where from the GSC to the grid is positive.

$$\frac{1}{2}C\frac{dV_{dc}^2}{dt} = -P_r - P_g = -\frac{3}{2}\hat{e}i_{qg} - P_r$$ (3.49)

Assuming that the DC-link voltage's variation is small, we have

$$CV_{dc0}\frac{dV_{dc}}{dt} = -\frac{3}{2}\hat{e}i_{qg} - P_r \qquad (3.50)$$

It can be seen that the DC-link voltage can be controlled by adjusting the q-axis current. In this case, positive feedback control should be pursued.

3.3.2 Inner Current Control

The converter is connected to the PCC through an inductor L_g. This inductor includes the effect of a filter and/or a transformer. The GSC output voltage, GSC current, and the coupling point voltage have the following relationship expressed in space vector and complex vector.

$$\overrightarrow{v_g} = L\frac{d\overrightarrow{i}}{dt} + \overrightarrow{e} \qquad (3.51)$$

$$\overline{V_g} = L\frac{d\overline{I}}{dt} + j\omega L_g\overline{I} + \overline{E} \qquad (3.52)$$

Align the reference frame's q-axis along with the coupling point voltage $\overrightarrow{e}$, we have

$$v_{qg} = L_g\frac{di_{qg}}{dt} + \omega L_g i_{dg} + e_q \qquad (3.53)$$

$$v_{dg} = L_g\frac{di_{dg}}{dt} - \omega L_g i_{qg} \qquad (3.54)$$

We can design feedback controllers based on virtual plant inputs $u_{qg} = v_{qg} - \omega L_g i_{dg} - e_q$ and $u_{dg} = v_{dg} + \omega L_g i_{qg}$. The feedback controller has the input from current measurement and generate the desired output u_{qg} and u_{dg}. Through feedforward of cross coupling items, the desired converter voltages can be found.

The overall GSC control is presented in Fig. 3.5. The block LPF stands for a low pass filter which gets rid of high-frequency noise to make the signal smooth. In GSC control, the coupling point voltage is same as the stator voltage.

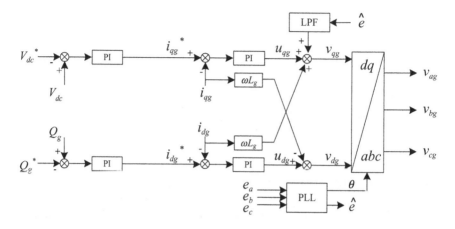

Figure 3.5 DFIG GSC control block diagram.

3.4 COMPLETE DFIG MODELING BLOCKS

RSC and GSC can be considered as two controllable voltage sources. These two voltage sources can be expressed in a reference frame where the stator voltage space vector is aligned with the q-axis. Note that the reference frames for DFIG model, RSC control, and GSC control are all aligned with the stator voltage. The RSC and GSC voltages are generated through the afore-mentioned control blocks. In Matlab/Simulink, feedback control blocks can be built. The converter controls can then be integrated with the DFIG model in the same dq reference frame.

There is one relationship not modeled yet: the DC-link capacitor dynamics or the relationship between the RSC and the GSC. The DC-link capacitor dynamics has to be considered as well. The overall dynamic model block diagram is shown in Fig. 3.6.

If a DFIG's stator voltage is assumed to be constant, then in the simulation block, v_{qs} can be assumed to be a constant. If the DFIG is connected to an infinity bus through a transmission line which can be considered as series RL components, the transmission line can be regarded as the additional stator resistance and stator leakage inductance. The stator voltage is given from the infinity bus voltage. Should the dynamics of the transmission line are more complicated, then the transmission line has to be adequately modeled. In Chapter 5, dynamics of a transmission line will be modeled and the integrated system model is presented. In Chapter 7, a DFIG

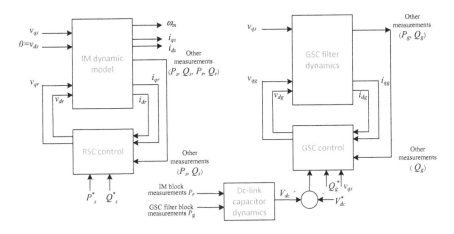

Figure 3.6 Overall DFIG dynamic model block diagram.

is connected to a power grid with multiple synchronous generators. The line dynamics are ignored.

3.5 EXAMPLES

3.5.1 Example 1: PSCAD Simulation of a Two-Level VSC with Sine PWM

In this example, we show PSCAD simulation results of a two-level VSC with sine PWM. The system is shown in Fig. 3.7. The DC side consists of two DC voltage source, each at 100 kV. The DC side is serving an RL three-phase load through a two-level VSC. Six gate signals will be generated through PWM. The phase A voltage (against the ground) v_a, per-phase voltage (E_a, against the neutral point o), and the neutral voltage v_o will be measured.

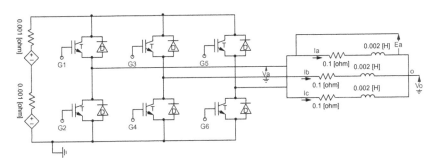

Figure 3.7 A two-level VSC system. DC voltage source: 100 kV.

Figure 3.8 shows the PMW scheme to generate six gate signals. First, three sinusoidal reference signals (magnitude 1, frequency 50 Hz, 120° apart from each other) are generated. The reference signal is then compared with the triangular carrier signal (amplitude 1, frequency 750 Hz). When the reference signal is greater than the carrier signal, the gate signal G1 is 1, G2 is 0. Otherwise, G1 is 0 and G2 is 1. The resulting phase A voltage to the ground is the amplification of the gate signal G1. v_a has two levels: 0 or 200 kV (the DC voltage) as shown in Fig. 3.9. The resulting A to neutral point voltage E_a has five levels as shown in Fig. 3.9. The fundamental component (or the signal after a LPF) is also shown in the first subplot. If we change the frequency and magnitude of the reference signal, we obtain

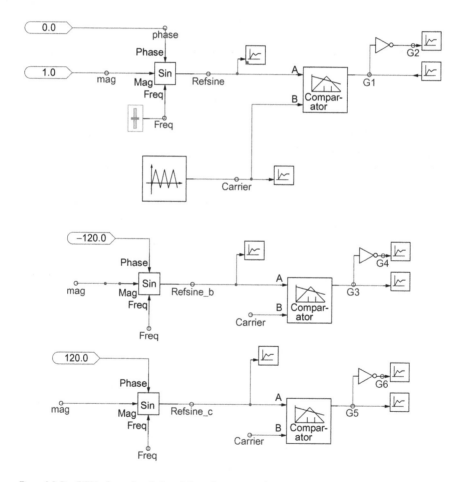

Figure 3.8 Sine PWM schemes for a balanced three-phase output voltage. Modulation index: 1.

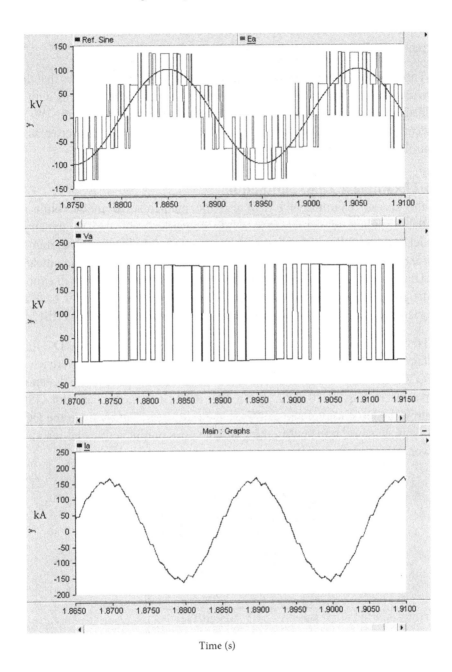

Figure 3.9 Measured signals.

converter voltage with the updated frequency and magnitude. The phase current i_a is shown as almost sinusoidal.

This example gives the PWM switching details and demonstrates that the converter output voltages can be considered as sinusoidal voltages with a fundamental frequency, e.g., 50 Hz. In the mathematical models, the PWM switching details will not be included. Instead, we treat a converter as a controllable three-phase AC voltage source.

3.5.2 Example 2: DFIG Simulation

In the second example, we will use the mathematical model built in Matlab/Simulink to demonstrate DFIG converter control. The dynamic model building follows the diagram shown in Fig. 3.6. The DFIG's parameters are listed in Table 3.1. The slip of the machine is 0.05 pu, which means the DFIG is running below the synchronous speed. Active power will export to the grid through the stator side. However, active power will flow back to the machine rotor ($P_r > 0$, $P_g < 0$) from the grid through GSC and RSC.

Here we assume the stator voltage is constant. At $t = 0.3$ s, the DC-link voltage reference will have a step change. The initial DC voltage reference is 1200 V. After 0.3 s, it becomes 1220 V. We will observe the system dynamic responses in Figs. 3.10–3.12. From Fig. 3.10, it is clear that with GSC control, the DC-link voltage can track the reference. Transients in the DC-link voltage cause transient in the grid converter output active and reactive power.

Table 3.1 Parameters of a Single 2 MW DFIG and the Aggregated DFIG in Network System	
Rated power	2 MW
Rated voltage	690 V
X_{ls}	0.09231 pu
X_M	3.95279 pu
X_{lr}	0.09955 pu
R_s	0.00488 pu
R_r'	0.00549 pu
H	3.5 s
X_g	0.3 pu (0.189 mH)
DC-link capacitor C	14,000 μF
DC-link rated voltage	1200 V

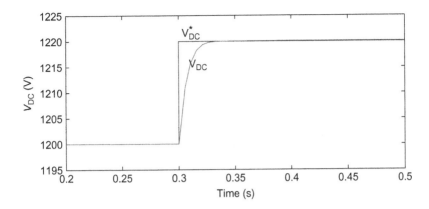

Figure 3.10 DC-Link voltage.

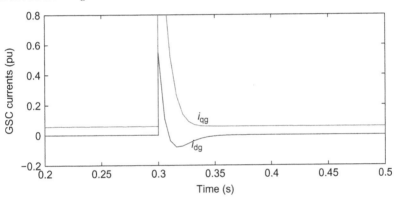

Figure 3.11 GSC qd-axis currents.

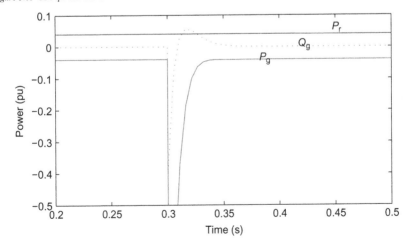

Figure 3.12 Real power from RSC to machine, real and reactive power from GSC to grid.

REFERENCES

[1] B.K. Bose, Modern Power Electronics and AC Drives. Prentice Hall, Upper Saddle River, NJ, 2001.

[2] S.R. Sanders, J.M. Noworolski, X.Z. Liu, G.C. Verghese, Generalized averaging method for power conversion circuits, IEEE Trans. Power Electron. 6(2) (1991) 251-259.

[3] R. Pena, J. Clare, G. Asher, Doubly fed induction generator using back-to-back pwm converters and its application to variable-speed wind-energy generation, IEEE Proc. Electr. Power Appl. 143(3) (1996) 231-241.

[4] A. Yazdani, R. Iravani, Voltage-Sourced Converters in Power Systems: Modeling, Control, and Applications, John Wiley & Sons, New York, 2010.

CHAPTER 4

Analysis of DFIG with Unbalanced Stator Voltage

4.1 STEADY-STATE HARMONIC ANALYSIS OF A DFIG 55

4.1.1 Steady-State Equivalent Circuit of a DFIG 56

4.1.2 Harmonic Components in Stator and Rotor Currents......... 58

4.1.3 Harmonic Components and Magnitudes of
Electromagnetic Torque 59

4.1.4 Example 61

4.2 UNBALANCED STATOR VOLTAGE DROP TRANSIENT
ANALYSIS 64

4.3 CONVERTER CONTROL TO MITIGATE UNBALANCE EFFECT 67

4.3.1 Negative Sequence Compensation via GSC..................... 67

4.3.1.1 Drawbacks of GSC Compensation 69

4.3.2 Negative Sequence Compensation via RSC 69

4.3.2.1 Dual-Sequence RSC Control 71

4.3.2.2 Proportional Resonant RSC Control 71

4.3.2.3 Drawbacks of RSC Compensation 73

REFERENCES... 73

This chapter presents DFIG analysis during unbalanced stator voltage conditions. The complexity resides in the rotor circuits. In addition, ripples will arise in torque and power. This chapter starts from steady-state circuit analysis, then proceeds to analyze the transient phenomena, and finally presents converter control to mitigate unbalance effect.

4.1 STEADY-STATE HARMONIC ANALYSIS OF A DFIG

The purpose of the analysis is to investigate the DFIG operation at unbalanced stator conditions and study the waveforms of the rotor currents and the electromagnetic torque. It is assumed that sinusoidal voltages are injected into the rotor and that the rotor injection voltage magnitude is constant

Modeling and Analysis of Doubly Fed Induction Generator Wind Energy Systems.
http://dx.doi.org/10.1016/B978-0-12-802969-5.00004-4

during the system disturbance. This assumption simplifies the RSC as a fixed voltage source.

There are two steps in the analysis. The first step is to identify the harmonic components in the rotor currents and the electromagnetic torque. The second step is to estimate the magnitude of each harmonic component.

4.1.1 Steady-State Equivalent Circuit of a DFIG

For analysis, the per-phase steady state equivalent circuit of a DFIG based on [1, 2] is shown in Fig. 4.1. Here N, ω_s and slip are defined based on sequence and harmonic conditions. For example, when $N = 1$, $\omega_s = \omega_e$ and slip $= s$, the circuit corresponds to the well-known positive sequence equivalent circuit of an induction machine.

Remarks.

- positive sequence circuit:

$$\begin{cases} N = 1, \\ \omega_s = \omega_e, \\ \text{slip} = s. \end{cases}$$

- negative sequence circuit:

$$\begin{cases} N = -1, \\ \omega_s = \omega_e, \\ \text{slip} = 2 - s. \end{cases}$$

The derivation of the steady-state circuit for the negative sequence components is given as follows. The equivalent circuit is derived by establishing the relationship of the voltages and currents expressed in qd-variables and further in phasors.

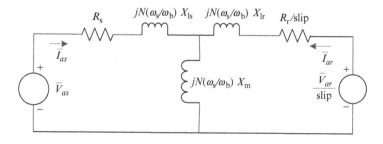

Figure 4.1 Steady-state induction machine circuit representation.

For the negative sequence, the q-axis and d-axis variables become DC variables at steady state when the reference frame rotates at a frequency of $-\omega_e$. The derivatives of the flux linkages are zero at steady state. Hence the voltage and current relationship is expressed in qd as

$$v_{qs}^{-e} = R_s i_{qs}^{-e} - \omega_e \psi_{ds}^{-e} \tag{4.1}$$

$$v_{ds}^{-e} = R_s i_{ds}^{-e} + \omega_e \psi_{qs}^{-e} \tag{4.2}$$

$$v_{qr}^{-e} = R_r i_{qr}^{-e} + (-\omega_e - \omega_m) \psi_{dr}^{-e} \tag{4.3}$$

$$v_{dr}^{-e} = R_r i_{dr}^{-e} - (-\omega_e - \omega_m) \psi_{qr}^{-e} \tag{4.4}$$

The relationship between a phasor $\tilde{F}_{as}$ at a given frequency and the corresponding qd variables in the reference frame rotating at the same frequency can be expressed as:

$$\sqrt{2}\tilde{F}_a = F_q - jF_d \tag{4.5}$$

where F can be voltages, currents, or flux linkages in the stator or rotor circuits.

Therefore, the stator and rotor voltage, current, and flux linkage relationship can be expressed in phasor form as

$$\tilde{V}_{as}^{-e} = R_s \tilde{I}_{as}^{-e} - j\omega_e \tilde{\psi}_{as}^{-e} \tag{4.6}$$

$$\tilde{V}_{ar}^{-e} = R_r \tilde{I}_{ar}^{-e} - j(\omega_e + \omega_m) \tilde{\psi}_{ar}^{-e} \tag{4.7}$$

The rotor relationship can be further expressed as

$$\frac{\tilde{V}_{ar}^{-e}}{2-s} = \frac{R_r}{2-s}\tilde{I}_{ar}^{-e} - j\omega_e \tilde{\psi}_{ar}^{-e}. \tag{4.8}$$

The equivalent circuit in Fig. 4.1 has $N = -1$, $\omega_s = \omega_e$ and slip $= 2 - s$. If the rotor voltage injection is assumed to be a balanced sinusoidal three-phase set, then the negative component $\tilde{V}_{ar}^{-e} = 0$.

Thus, the stator and rotor currents are induced by both the positive sequence voltages and negative sequence voltages. The rotor currents have two components, one at the low frequency $s\omega_e$ having a root mean square (RMS) magnitude of I_{as+} and the other at the high frequency $(2 - s)\omega_e$ with a magnitude of I_{as-}.

4.1.2 Harmonic Components in Stator and Rotor Currents

The stator frequency is assumed to be 60 Hz. During stator unbalance, the magnitudes of the three phase voltages will not be the same. Also the phase angle displacements of the three voltages will not be 120°. Using symmetric component theory, any three-phase voltages can be decomposed into a positive-, a negative-, and a zero-sequence component. The stator currents will in turn have positive-, negative-, and zero-sequence components.

For an induction machine, the sum of the rotor injection frequency and the rotor rotating frequency equals to the stator frequency or $\omega_r + \omega_m = \omega_s$. For the positive sequence voltage set with frequency ω_s applied to the stator side, the resulting rotor currents have a frequency $\omega_r = \omega_s - \omega_m = s\omega_s$, or slip frequency ω_{sl}.

The negative sequence voltage set can be seen as a three-phase balanced set with a negative frequency $-\omega_s$. Thus the induced flux linkage in rotor circuit and the rotor currents have a frequency of $-\omega_s - \omega_m = -(2 - s)\omega_s$.

Observed from the synchronous reference frame $qd+$ with a rotating speed ω_e, the first component (positive sequence) has a frequency of $s\omega_s - (\omega_e - \omega_m) = 0$, or a DC component, and the second component has a frequency of $-(2 - s)\omega_s - (\omega_e - \omega_m) = -2\omega_e$, i.e., 120 Hz. Observed from the negative synchronous reference frame $qd-$ which rotates clockwisely with the synchronous speed ω_e, then the positive sequence component has a frequency of $s\omega_s - (-\omega_e - \omega_m) = 2\omega_e$, and the negative sequence component has a frequency of $-(2 - s)\omega_s - (-\omega_e - \omega_m) = 0$. The two reference frames are shown in Fig. 4.2 and Table 4.1 shows the components of the rotor currents in abc and the two reference frames.

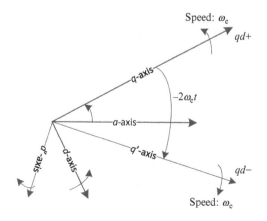

Figure 4.2 The two reference frames: synchronous and negatively synchronous.

Table 4.1 Rotor Current Components Observed in Various Reference Frames			
	abc	qd^+	qd^-
Positive	$s\omega_e$	0	$2\omega_e$
Negative	$-(2-s)\omega_e$	$-2\omega_e$	0

The rotor currents in both reference frames will have a DC component and a high frequency component. To extract the harmonic components in the rotor currents, both a synchronous reference frame qd^+ and a clockwise synchronously rotating reference frame qd^- (Fig. 4.2) will be used. A LPF with a suitable cutoff frequency can be used to extract the DC components which correspond to the magnitudes of the two harmonic components. The scheme for extracting the DC components is shown in Fig. 4.3.

4.1.3 Harmonic Components and Magnitudes of Electromagnetic Torque

The zero sequence stator currents will not induce a torque [2]. Meanwhile, the 0-axis stator circuit and 0-axis rotor circuit are completely decoupled. Hence, the 0-axis variable transformed from stator side will not induce any voltage at rotor side. In most machines, wye connection is used so even the stator side will have no zero sequence currents.

Under unbalanced stator condition, the stator current has two components: positive sequence $\bar{I}_{s+}$ and negative sequence components $\bar{I}_{s-}$. The rotor current also has two components: positive sequence $\bar{I}_{r+}$ and negative sequence components $\bar{I}_{r-}$. The electromagnetic torque is produced by the interactions between the stator and rotor currents. The torque can be decomposed into four components:

$$T_e = T_{e1} + T_{e2} + T_{e3} + T_{e4} \tag{4.9}$$

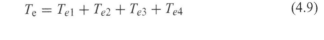
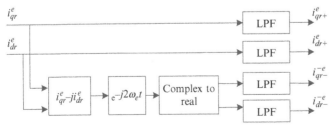

Figure 4.3 Scheme for extracting DC components.

60   Modeling and Analysis of Doubly Fed Induction Generator Wind Energy Systems

where T_{e1} is due to the interaction of $\bar{I}_{s+}$ and $\bar{I}_{r+}$, T_{e2} is due to the interaction of $\bar{I}_{s-}$ and $\bar{I}_{r-}$, T_{e3} is due to the interaction of $\bar{I}_{s+}$ and $\bar{I}_{r-}$, and T_{e4} is due to the interaction of $\bar{I}_{s-}$ and $\bar{I}_{r+}$.

It will be convenient to use both the synchronously rotating reference frame and the negative synchronous reference frame to compute T_e. For example, T_{e1} can be identified as a DC variable. T_{e2} can be identified as a DC variable. T_{e3} and T_{e4} are pulsating torques with a frequency $2\omega_e$. The expressions for the torque components are as follows:

$$T_{e1} = 3\left(\frac{P}{2}\right)X_m\mathcal{I}(\bar{I}^e_{s+}\bar{I}^{*e}_{r+}) \tag{4.10}$$

$$T_{e2} = 3\left(\frac{P}{2}\right)X_m\mathcal{I}[\bar{I}^{-e}_{s-}\bar{I}^{*-e}_{r-}] \tag{4.11}$$

$$T_{e3} = 3\left(\frac{P}{2}\right)X_m\mathcal{I}[\bar{I}^e_{s+}\bar{I}^{*e}_{r-}] \tag{4.12}$$

$$= 3\left(\frac{P}{2}\right)X_m\mathcal{I}[\bar{I}^e_{s+}\bar{I}^{*-e}_{r-}e^{j2\omega_e t}] \tag{4.13}$$

$$T_{e4} = 3\left(\frac{P}{2}\right)X_m\mathcal{I}[\bar{I}^e_{s-}\bar{I}^{*e}_{r+}] \tag{4.14}$$

$$= 3\left(\frac{P}{2}\right)X_m\mathcal{I}[\bar{I}^{-e}_{s-}\bar{I}^{*e}_{r+}e^{-j2\omega_e t}] \tag{4.15}$$

where $\bar{I}_s = 1/\sqrt{2}(i_{qs} - ji_{ds})$ and $\bar{I}_r = 1/\sqrt{2}(i_{qr} - ji_{dr})$, and F^e_+ is the qd variables of the positive sequence component in synchronous reference frame; F^e_- is the qd variables of the negative sequence component in synchronous reference frame; F^{-e}_+ is the qd variables of the positive sequence component in negative synchronous reference frame; F^{-e}_- is the qd variables of the positive sequence component in negative synchronous reference frame.

The torque expression under unbalanced stator condition is

$$T_e = T_{e0} + T_{esin2}\cdot\sin(2\omega_s t) + T_{ecos2}\cdot\cos(2\omega_s t) \tag{4.16}$$

where the expression of T_{e0}, T_{esin2} and T_{ecos2} can be found in (4.17).

$$
\begin{bmatrix} T_{e0} \\ T_{esin2} \\ T_{ecos2} \end{bmatrix} = \frac{3PX_{\mathrm{m}}}{4} \begin{bmatrix} i^e_{qs+} & -i^e_{ds+} & i^{-e}_{qs-} & -i^{-e}_{ds-} \\ i^{-e}_{ds-} & i^{-e}_{qs-} & -i^e_{ds+} & -i^e_{qs+} \\ i^{-e}_{qs-} & -i^{-e}_{ds-} & i^e_{qs+} & -i^e_{ds+} \end{bmatrix} \begin{bmatrix} i^e_{dr+} \\ i^e_{qr+} \\ i^{-e}_{dr-} \\ i^{-e}_{qr-} \end{bmatrix} \quad (4.17)
$$

The harmonic components in the torque can be computed from positive and negative stator/ rotor currents.

4.1.4 Example
A 3HP DFIG is used for analysis and simulation. The machine parameters are shown in Table 4.2 as follows.

The initial condition of the machine is the stalling state. A balanced three-phase voltage and a mechanical torque 10 Nm are applied to the stator at $t = 0$ second. The system configuration is shown in Fig. 4.4. At $t = 1$ second, the voltage of phase A drops to zero. The fault is cleared at $t = 1.5$ second. The simulation is performed in Matlab/Simulink and the results are shown in Figs. 4.5–4.7.

Table 4.2 Induction Machine Parameters	
$R_s(\Omega)$	0.435
X_{ls} (Ω)	0.754
X_m (Ω)	26.13
X_{lr} (Ω)	0.754
r_r (Ω)	0.816
$J(kg.m^2)$	0.089

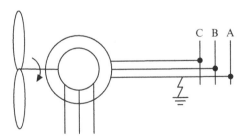

Figure 4.4 A grid-interconnected DFIG system configuration.

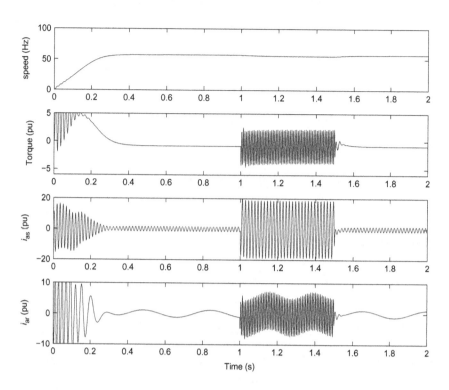

Figure 4.5 Dynamic responses of rotor speed, electromagnetic torque, phase a stator current, and phase a rotor current.

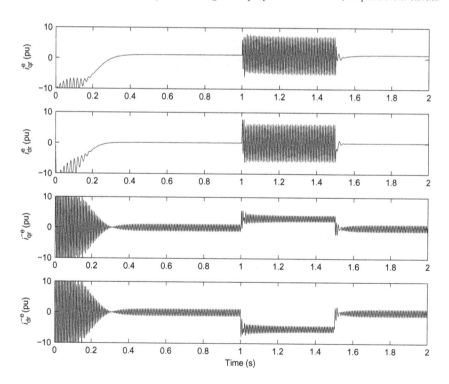

Figure 4.6 Dynamic response of i_{qr}^e, i_{dr}^e, i_{qr}^{-e}, and i_{dr}^{-e}.

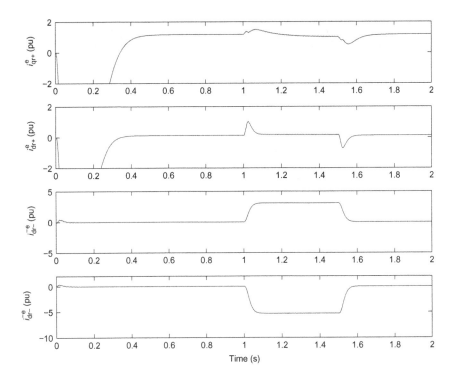

Figure 4.7 Harmonic components after extracting strategy from Fig. 4.3. (a) i^e_{qr+}—DC component of i_{qr} observed in the synchronously rotating reference frame, (b) i^e_{dr+}—DC component of i_{dr} observed in the synchronously rotating reference frame, (c) i^{-e}_{qr-}—DC component of i_{qr} observed in the negatively synchronously rotating reference frame, (d) i^{-e}_{dr-}—DC component of i_{dr} observed in the synchronously rotating reference frame.

Figure 4.5 shows the dynamic responses of the rotor speed, electromagnetic torque, and stator and rotor currents in phase A. Figure 4.6 shows the dynamic responses of the rotor currents in the synchronous reference frame and the negative synchronous reference frame. At the steady state balanced stator condition, the rotor speed is equivalent to 57.2 Hz while the slip frequency of the rotor currents is 2.8 Hz. The two add up to 60 Hz. During unbalance, it is found that the torque has a 120 Hz pulsating component. It is seen from the simulation plots that the rotor current consists of two components: a low frequency component and a high frequency component. According to the analysis the two frequencies are the slip frequency and a frequency close to 120 Hz ($(2 - s)\omega_e$). The rotor currents observed in the synchronous reference frame have a DC component and a 120 Hz component.

Table 4.3 Calculated Sequence Components in Stator Voltages, Stator Currents and Rotor Currents Assuming Slip = 0.12

Voltage	Mag	Current	Mag	Current	Mag	Frequency
$\lvert V_s^+ \rvert$	88.53 V	$\lvert I_s^+ \rvert$	4.78 A	$\lvert I_r^+ \rvert$	4.16 A	$s\omega_e$
$\lvert V_s^- \rvert$	44.26 V	$\lvert I_s^- \rvert$	25.87 A	$\lvert I_r^- \rvert$	25.14 A	$(2-s)\omega_e$
$\lvert V_s^0 \rvert$	44.26 V	$\lvert I_s^0 \rvert$	50.85 A	$\lvert I_r^0 \rvert$	0 A	0

Table 4.4 Harmonic Components in the Rotor Currents and the Electromagnetic Torque from Simulation and Analysis During Unbalanced Condition (Slip = 4.5/60)

	Simulation	Analysis
i_{qr+}^e	6 A	5.8 A
i_{dr+}^e	0.9 A	0.94 A
$\lvert I_{r+} \rvert$	4.29 A	4.16 A
i_{qr-}^{-e}	17.9 A	17.8 A
i_{dr-}^{-e}	-30.8 A	-30.8 A
$\lvert I_{r-} \rvert$	25.2 A	25.15 A
T_{e0}	-10 N m	-9.89 N m
$\sqrt{T_{esin2}^2 + T_{ecos2}^2}$	35 N m	36.3 N m

The magnitudes of the harmonic components can be computed from the equivalent phasor circuits in Fig. 4.1. Table 4.3 lists the calculated positive, negative, and zero sequence components in the stator voltage, stator current, and rotor current.

The sequence components of the stator voltage, the stator currents, and the rotor currents are listed in Table 4.3. A comparison of the analysis results from the circuit in Fig. 4.1 and simulation results shows that the rotor current and torque components from analysis agree with the simulation results in Table 4.4.

4.2 UNBALANCED STATOR VOLTAGE DROP TRANSIENT ANALYSIS

The transients in rotor voltages due to stator voltage dip are analyzed in [3]. In this section, we use an example to explain the effect of unbalanced stator voltage on the rotor circuit transients.

Example. Express the stator flux linkage when the stator voltage has an unsymmetrical voltage drop.

Solution. Recall Chapter 2 (Example 2) of examining rotor voltage due to stator voltage symmetric drop. Here we again use the space vector model to solve this problem.

The key dynamic equation that relates stator voltage and stator flux linkage is the following equation:

$$\overrightarrow{v_s} = R_s \overrightarrow{i_s} + \frac{d\overrightarrow{\psi_s}}{dt} \tag{4.18}$$

$$\text{where } \overrightarrow{\psi_s} = L_m(\overrightarrow{i_s} + \overrightarrow{i_r}) + L_{ls}\overrightarrow{i_s} \tag{4.19}$$

With the assumption of zero rotor current, the stator flux is only related to the stator current.

$$\overrightarrow{\psi_s} \approx L_s \overrightarrow{i_s} \tag{4.20}$$

Therefore, we have

$$\overrightarrow{v_s} \approx \frac{R_s}{L_s}\overrightarrow{\psi_s} + \frac{d\overrightarrow{\psi_s}}{dt}. \tag{4.21}$$

Let us first consider a symmetric stator voltage. Example 2 in Chapter 2 examined a case of stator voltage dropping to zero. Here, we will consider a more general case where the stator voltage initially has a magnitude of V_1 and after the drop a magnitude of V_2. The amplitude of the voltage waveform, which is the same as the magnitude of the space vector, is assumed to be V_2.

At $t = t_0^-$, the space vector of the stator voltage is $\overrightarrow{v}_s(t) = V_1 e^{j\omega_s t}$. At $t = t_0^+$, the space vector of the stator voltage becomes $\overrightarrow{v}_s(t) = V_2 e^{j\omega_s t}$.

For $t < t_0^-$, let us find the steady-state solution of the stator flux linkage $\overrightarrow{\psi_s}(t)$. We know that it should have a format of $\bar{K}e^{j\omega_s t}$, where $\bar{K}$ is a complex vector. Substituting $\overrightarrow{\psi_s}$ by $\bar{K}e^{j\omega_s t}$ in (4.21), we have

$$\bar{K}_s\left(\frac{R_s}{L_s} + j\omega_s\right)e^{j\omega_s t} = V_1 e^{j\omega_s t}$$

$$\bar{K} = \frac{V_1}{\frac{R_s}{L_s} + j\omega_s} \tag{4.22}$$

If we ignore $R_s \approx 0$, then $\bar{K} = \frac{V_1}{j\omega_s}$. Therefore, $\overrightarrow{\psi_s}(t_0^-) = \frac{V_1}{j\omega_s}e^{j\omega_s t_0}$. Since flux linkages cannot change abruptly, therefore, $\overrightarrow{\psi_s}(t_0^+) = \overrightarrow{\psi_s}(t_0^-)$.

Then for $t \geq t_0^+$, let us find the time-domain expression of the flux linkage. The dynamic relationship between the stator voltage and the stator flux linkage is as follows:

$$V_2 e^{j\omega_s t} = \frac{R_s}{L_s} \overrightarrow{\psi_s} + \frac{d\overrightarrow{\psi_s}}{dt} \tag{4.23}$$

The expression of the stator flux linkage consists of a steady-state term (forced flux) and a homogeneous term (natural flux). The forced flux can be found as

$$\overrightarrow{\psi}_{s,F}(t) = \frac{V_2}{\frac{R_s}{L_s} + j\omega_s} e^{j\omega_s(t-t_0)} \approx \frac{V_2}{j\omega_s} e^{j\omega_s(t-t_0)}. \tag{4.24}$$

The nature flux has the form of

$$\overrightarrow{\psi}_{s,N}(t) = K e^{\frac{-R_s}{L_s}(t-t_0)}, \tag{4.25}$$

where K is a coefficient.

Using the initial condition that $\overrightarrow{\psi}_s(t_0^+) = \frac{V_1}{j\omega_s} e^{j\omega_s t_0}$, we can find $K = \frac{V_1 - V_2}{j\omega_s} e^{j\omega_s t_0}$.

Therefore, for a symmetric stator voltage drop, the time-domain stator flux linkage has an expression as follow:

$$\overrightarrow{\psi}_s(t) \approx \frac{V_2}{j\omega_s} e^{j\omega_s t} + \frac{V_1 - V_2}{j\omega_s} e^{j\omega_s t_0} e^{\frac{-R_s}{L_s}(t-t_0)} \tag{4.26}$$

We now proceed to investigate the stator flux linkage expression for a non-symmetric stator voltage drop. A non-symmetric three-phase voltage can be decomposed into positive-, negative-, and zero-sequence components. Note that the corresponding space vector for the zero-sequence component is zero. Therefore, the space vector for the stator voltage can be expressed in the following:

$$\overrightarrow{v}_s(t) = \begin{cases} V^+ e^{j\omega_s t} + V^- e^{-j\omega_s t}, & \text{for } t \geq t_0^+ \\ V_s e^{j\omega_s t}, & \text{for } t \leq t_0^- \end{cases} \tag{4.27}$$

The forced flux linkage due to the positive sequence component of the stator voltage is $\frac{V^+}{j\omega_s} e^{j\omega_s t}$. The forced flux linkage due to the negative sequence component is $\frac{V^-}{-j\omega_s} e^{-j\omega_s t}$. And the natural flux linkage is $K e^{\frac{-R_s}{L_s}(t-t_0)}$.

Using the initial condition, we can find

$$K = \frac{V_s - V^+}{j\omega_s}e^{j\omega_s t_0} + \frac{V^-}{j\omega_s}e^{-j\omega_s t_0}. \qquad (4.28)$$

The complete expression of the stator flux linkage is presented as follows:

$$\overrightarrow{\psi_s}(t) = \frac{V^+}{j\omega_s}e^{j\omega_s t} + \frac{V^-}{-j\omega_s}e^{-j\omega_s t} + \left(\frac{V_s - V^+}{j\omega_s}e^{j\omega_s t_0} + \frac{V^-}{j\omega_s}e^{-j\omega_s t_0}\right)e^{\frac{-R_s}{L_s}(t-t_0)}$$

$$(4.29)$$

4.3 CONVERTER CONTROL TO MITIGATE UNBALANCE EFFECT

Transients in rotor voltages due to unsymmetrical stator voltage dip have been analyzed in [3]. Section 4.2 provides a brief analysis on its effect in stator flux. Suitably sized converters can accommodate transients. On the other hand, the more severe operation problems are the torque ripples and the DC-link voltage ripples due to the negative sequence components in the stator and rotor currents [4]. The steady-state analysis of ripples in rotor currents and electromagnetic torque is presented in Section 4.1. In summary, due to unbalanced stator voltage conditions, negative-sequence components in stator currents induce a high frequency component ($\omega_e + \omega_m$) or ($2 - s)\omega_e$ in rotor currents and pulsations at $2\omega_e$ frequency in electromagnetic torques. In this section, negative sequence compensation techniques via RSC and GSC are presented.

4.3.1 Negative Sequence Compensation via GSC
Negative sequence compensation via GSC is presented in [5]. The philosophy is to let the GSCs compensate the negative sequence currents required in the network during any unbalanced operation. The circuit model is shown in Fig. 4.8. The GSCs will supply the negative sequence current components to the grid. Hence the stator currents will remain balanced. With balanced stator currents, a rotating magnetic field will be formed in the air gap, which induces balanced EMF in the rotor circuits. The rotor currents will remain balanced. Therefore, the torque will remain ripple free.

For negative sequence compensation via GSC, the current controllers of the GSC will measure the network currents, extract the negative sequence components and generate the required negative sequence currents from the GSC for compensation. The reference values of the negative sequence

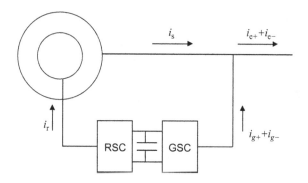

Figure 4.8 The control philosophy: negative sequence compensation through GSC: $i_{g-} = i_{e-}$. Stator current i_s is free of negative sequence components.

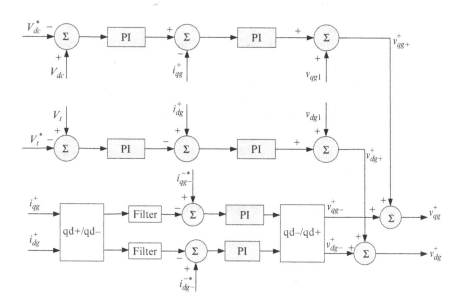

Figure 4.9 Dual sequence control loops of GSC. $v_{qg1} = v_{qs} + \omega_e L_g i_{dg}$; $v_{dg1} = v_{ds} - \omega_e L_g i_{dg}$. L_g is the inductance of the transformer connecting the GSC to the grid.

currents come from the measurements of the currents to the grid $i_{e,abc}$. The negative sequence components of $i_{e,abc}$ are then extracted through abc/qd^- transformation and LPFs.

Figure 4.9 presents the dual sequence control loops of GSC. The positive-sequence control loop is the same as the control loop presented in Chapter 3. In addition, the negative-sequence control loop generates negative-sequence voltage. The assumption of Fig. 4.9 is that all measurements are available in the synchronous reference frame $dq+$. For the negative-sequence control, the currents are first been transformed in the dq-reference frame. In that reference frame, the negative-sequence current component appears as DC variables while the positive-sequence component appears as 120 Hz ripple. With a LPF, only the negative-sequence current component (i_{qg}^-, i_{dg}^-) will be obtained. For the given reference values, PI controllers are suitable to the measurements to track the reference values. The output from the PI controllers are negative-sequence voltage in $dq-$ frame. This component should be transformed back to $dq+$ and added back to the voltage generated by the positive-sequence control.

4.3.1.1 Drawbacks of GSC Compensation

Since the GSC compensates a negative sequence current to the grid, the three-phase voltage from the GSC should provide the negative sequence component as well. The instantaneous power through the GSC will have pulsating components. The dynamic equation of DC-ink voltage is given by:

$$CV_{DC}\frac{dV_{DC}}{dt} = -P_g - P_r \tag{4.30}$$

where P_g and P_r are the GSC and RSC instantaneous powers. The directions of the powers are determined by the direction of the currents in Fig. 4.8.

The detailed analysis of the DC-link voltage due to unbalanced grid voltages can be found in [6]. If there is only negative sequence compensation from the GSC, and with the assumption that the rotor power P_r has only DC component, the DC-link voltage will have ripples with two pulsating components at frequencies of $2\omega_e$ and $4\omega_e$ [6]. The more unbalanced the grid voltage, the higher the magnitude of the pulsating power, and hence the higher the magnitude of the DC-link voltage ripple.

4.3.2 Negative Sequence Compensation via RSC

Torque pulsation can be eliminated by negative sequence compensation via the RSC [7, 8]. The steady-state negative sequence circuit model with the RSC compensation has been derived in Section 4.1 (Fig. 4.1). From the circuit model, it can be seen that for a given negative-sequence stator voltage, a negative sequence rotor voltage generated the RSC has the potential to eliminate the negative sequence rotor current ($\bar{I}_{ar}^- = 0$) or the negative sequence stator current ($\bar{I}_{as}^- = 0$) or the torque pulsation. The derivation of the rotor current reference values to eliminate the torque pulsation can be found in [7]. In the following paragraphs, torque pulsation elimination procedure will be described.

The control objective is to eliminate the ripples in the torque and therefore the reference values of the negative sequence rotor currents need to be calculated. The electromagnetic torque can be expressed in the form of the stator flux linkage and the rotor current [2]:

$$T_e = \frac{3}{2}\frac{P}{2}\frac{L_m}{L_s}\Re(-j\vec{\psi}_s\,\vec{i}_r^{\,*})\qquad(4.31)$$

The rotor current space vector and the stator flux linkage space vector in qd^+ reference frame consisting of positive and negative sequence components can be expressed as follows:

$$\begin{cases}\vec{I_r}^+ = \overline{I}_{r+}^+ + \overline{I}_{r-}^-e^{-j2\omega_e t}\\[4pt]\vec{\psi_s}^+ = \overline{\psi}_{s+}^+ + \overline{\psi}_{s-}^-e^{-j2\omega_e t}.\end{cases}\qquad(4.32)$$

Therefore, the electromagnetic torque in (4.31) has three components: $T_e^+ = T_{edc}^+ + T_{ecos}^+\cos(2\omega_e t) + T_{esin}^+\sin(2\omega_e t),$

$$\text{where }\begin{cases}T_{edc}^+ = K(\psi_{qs+}^+ i_{qr+}^+ + \psi_{ds+}^+ i_{dr+}^+ + \psi_{qs-}^- i_{qr-}^- + \psi_{ds-}^- i_{dr-}^-)\\[4pt]T_{ecos}^+ = K(\psi_{qs+}^+ i_{qr-}^- + \psi_{ds+}^+ i_{dr-}^- + \psi_{qs-}^- i_{qr+}^+ + \psi_{ds-}^- i_{dr+}^+)\\[4pt]T_{esin}^+ = K(\psi_{ds+}^+ i_{qr-}^- - \psi_{qs+}^+ i_{dr-}^- - \psi_{ds-}^- i_{qr+}^+ + \psi_{qs-}^- i_{dr+}^+)\end{cases}\qquad(4.33)$$

where $K = \frac{3}{2}\frac{P}{2}\frac{M}{L_s}$. To minimize ripples in the electromagnetic torque, the AC components of the torque should be set to zeros, that is,

$$\begin{cases}T_{ecos}^+ = 0\\[4pt]T_{esin}^+ = 0.\end{cases}\qquad(4.34)$$

From the above requirements, the reference values of the negative sequence rotor currents can be computed based on the reference positive sequence rotor currents and the stator flux linkage measurements. The PI controllers in the negative sequence control loops make sure the negative sequence components in the rotor currents track the referenced values.

Computation of the reference negative sequence currents requires extensive information of the stator flux linkage and the positive sequence reference rotor current. Time delay will be introduced in such control structure.

4.3.2.1 Dual-Sequence RSC Control

Dual sequence controllers to separately control the positive sequence and negative sequence rotor current applied in [4, 9] are shown in Fig. 4.10. Low pass filters are applied to separate the currents after abc/dq transformation. Filters can introduce time delay and deteriorate control performance. Therefore, in [10, 11], the main controller dealing with the positive sequence is implemented without filter, only the auxiliary controller dealing with the negative sequence rotor currents has filters. Note the negative-sequence rotor current commands are computed from (4.34).

4.3.2.2 Proportional Resonant RSC Control

Instead of realizing dual-sequence control in $dq+$ and $dq-$ reference frame, the control can be realized in one reference frame $\alpha\beta$ reference frame. In the $\alpha\beta$ reference frame, we deal with AC signals. Ac signal tracking can be realized by proportional resonant (PR) control. A PR controller can be considered as an AC signal tracker just as a PI controller is a DC signal tracker [12]. For example, a compensator with a transfer function $K_p + \frac{K_R s}{s^2 + \omega_c^2}$ can make sure the open-loop function has an infinitive magnitude at ω_e.

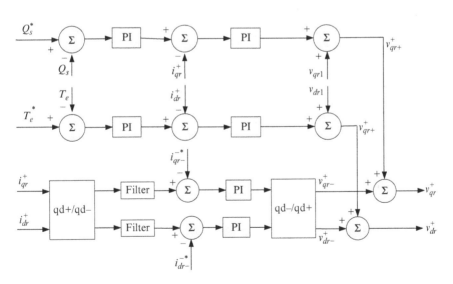

Figure 4.10 Dual sequence control loops for RSC. $v_{qr1} = s\omega_e\sigma L_r i_{dr} + s\omega_e\frac{L_m}{L_s}\psi_{ds}$, $v_{dr1} = -s\omega_e\sigma L_r i_{qr}$, where $\sigma = 1 - \frac{L_m^2}{L_s L_r}$.

Therefore, the close-loop system will realize zero error at that frequency. For a signal with a frequency of ω_e, that PR controller can make the signal track its reference.

In addition, instead of computing the reference values of the RSC negative sequence currents, the reference values for the negative sequence rotor currents can be set to zeros. This way, pulsations in both the rotor currents and the torque can be suppressed if not fully eliminated. Hence, the purpose of the negative sequence RSC current controller is to eliminate the negative sequence rotor currents and the reference negative sequence rotor currents are zero.

With such knowledge, the dual sequence control structure can be simplified using proportional resonance (PR) or proportional integral resonant (PIR) controller. A PR controller has been tested to eliminate the negative sequence rotor current by the authors in [13]. The PR-based control loops for the RSC is shown in Fig. 4.11.

Since the negative sequence current reference values are set to zeros, the reference currents in $\alpha\beta$ contain only positive sequence information. The measurements will also be transferred to $\alpha\beta$ frame. Rotor currents in $\alpha\beta$ will only have components with a frequency of ω_e under unbalanced grid voltage conditions. PR controllers are effective for AC signal tracking. Hence the controller will eliminate the negative sequence currents in the rotor circuits.

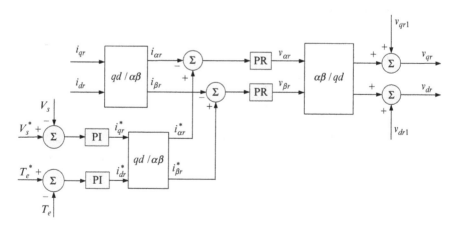

Figure 4.11 Proposed control loops for RSC. The transfer function in the PR block is $K_P + \frac{K_I s}{s^2 + \omega_e^2}$. $v_{qr1} = s\omega_e \sigma L_r i_{dr} + s\omega_e \frac{L_m}{L_s} \psi_{ds}$, $v_{dr1} = -s\omega_e \sigma L_r i_{qr}$.

4.3.2.3 Drawbacks of RSC Compensation

Without losing the generality, the RSC needs to inject a negative sequence voltage. Similar as the consequence of negative sequence compensation via GSC, the DC-link voltage will have ripples with pulsating components. Therefore using RSC negative sequence compensation alone leads to DC-link voltage ripples.

It seems that torque ripple suppression and DC-link voltage ripple suppression, cannot be achieved simultaneously using either RSC compensation or GSC compensation. Therefore both RSC and GSC controls are used in [4, 9–11]. Interested readers can consult these references for more information on RSC and GSC coordination under unbalanced conditions.

REFERENCES

[1] A. Fitzgerald, C. Kingsley, A. Kusko, Electric Machinery, McGraw-Hill Book Company, New York, 1971.

[2] P. Krause, Analysis of Electric Machinery, McGraw-Hill, New York, 1986.

[3] J. Lopez, E. Gubia, P. Sanchis, X. Roboam, L. Marroyo, Wind turbines based on doubly fed induction generator under asymmetrical voltage dips, IEEE Trans. Energy Convers. 23(1) (2008) 321-330.

[4] Y. Zhou, P. Bauer, J. Ferreira, J. Pierik, Operation of grid-connected DFIG under unbalanced grid voltage condition, IEEE Trans. Energy Convers. 24(1) (2009) 240-246.

[5] R. Pena, R. Cardenas, E. Escobar, Control system for unbalanced operation of stand-alone doubly fed induction generators, IEEE Trans. Energy Convers. 22(2) (2007) 544-545.

[6] X. Wu, S. Panda, J. Xu, Analysis of the instantaneous power flow for three-phase pwm boost rectifier under unbalanced supply voltage conditions, IEEE Trans. Ind. Electron. 23(4) (2008) 1679-1691.

[7] L. Xu, Y. Wang, Dynamic modeling and control of DFIG-based wind turbines under unbalanced network conditions, IEEE Trans. Power Syst. 22(1) (2007) 314-323.

[8] J. Hu, Y. He, Modeling and enhanced control of DFIG under unbalanced grid voltage conditions, Electric Power Syst. Res. 79(2) (2008) 273-281.

[9] O. Gomis-Bellmunt, A. Junyent-Ferre, A. Sumper, J. Bergas-Jane, Ride-through control of a doubly fed induction generator under unbalanced voltage sags, IEEE Trans. Energy Convers. 23(4) (2008) 1036-1045.

[10] L. Xu, Coordinated control of DFIG's rotor and grid side converters during network unbalance, IEEE Trans. Power Electron. 23(3) (2008) 1041-1049.

[11] L. Xu, Enhanced control and operation of DFIG-based wind farms during network unbalance, IEEE Trans. Energy Convers. 23(4) (2008) 1073-1081.

[12] R. Teodorescu, F. Blaabjerg, M. Liserre, P. Loh, Proportional-resonant controllers and filters for grid-connected voltage-source converters, IEEE Proc. Electr. Power Appl. 153(5) (2006) 750-762.

[13] L. Fan, R. Kavasseri, H. Yin, C. Zhu, M. Hu, Control of DFIG for rotor current harmonics elimination, in: Proceedings of IEEE Power & Energy Society General Meeting, Calgary, Canada, Jul. 2009.

CHAPTER 5

State-Space Based DFIG Wind Energy System Modeling

5.1 STATE-SPACE MODEL OF A SERIES COMPENSATED NETWORK **75**

5.2 STATE-SPACE MODEL OF DFIG WIND ENERGY SYSTEM **76**

5.2.1 Induction Generator Model **76**

5.2.2 DC-Link Model **76**

5.2.3 Torsional Dynamics Model **77**

5.2.4 DFIG Converter Controls **78**

5.3 INTEGRATED SYSTEM MODEL **78**

5.4 APPLICATION OF SSR ANALYSIS **80**

5.4.1 Introduction **80**

5.4.2 Analysis of SSR in a DFIG **82**

5.4.3 Impact of Wind Speed and Compensation Level on SSR **83**

5.4.4 Impact of DFIG Current Controllers on SSR **86**

5.4.5 Results of TI Effect **88**

5.4.5.1 Impact of Compensation Level on Torsional Mode **89**

5.4.5.2 Impact of Wind Speed on TI **90**

APPENDIX **92**

REFERENCES **93**

In this chapter, we will discuss how to build a state-space model for DFIG-based wind energy system. The chapter is organized as follows. The state-space models of a series-compensated transmission line and a DFIG wind energy system in the *dq* reference frame are first introduced, followed by the presentation on integrated system modeling techniques. The application of the state-space model is Type-3 wind energy system subsynchronous resonance (SSR) analysis.

Modeling and Analysis of Doubly Fed Induction Generator Wind Energy Systems.
http://dx.doi.org/10.1016/B978-0-12-802969-5.00005-6

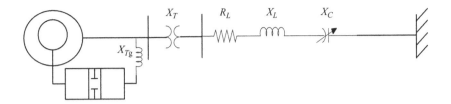

Figure 5.1 The study system.

The study system is shown in Fig. 5.1, where a DFIG-based wind farm (892.4 MVA from aggregation of 2 MW units) is connected to a series-compensated line whose parameters correspond to the IEEE first benchmark model for SSR studies [1]. The wind farms can be considered as coherent generators and can be represented by one large DFIG. This approach has been practiced in system studies [2].

5.1 STATE-SPACE MODEL OF A SERIES COMPENSATED NETWORK

In power system dynamic studies, transmission networks are often represented by phasor-based algebraic equations with dynamics ignored. This is not the case for SSR analysis, where dynamics in the transmission network is related to the phenomena. The transmission network is considered as a RLC circuit with lumped parameters. The dynamic equations can be expressed as

$$Ri_p + L\frac{di_p}{dt} + v_{c,p} = v_{t,p} - e_p$$

$$C\frac{dv_{c,p}}{dt} = i_p$$

where $p = a, b, c$, i_p is the phase current from the generator bus to the grid, $v_{t,p}$ is the phase voltage at the generator bus, $v_{c,p}$ is the voltage across the capacitor, and e_p is the voltage of the grid.

Synchronous reference frame is widely used for modeling induction machines [3]. The same reference frame should be adopted for the network model. In the synchronous reference frame, the dynamics of the series-compensated system can be described by the following equation in real values:

$$\frac{d}{dt}\begin{bmatrix} v_{cq} \\ v_{cd} \\ i_q \\ i_d \end{bmatrix} = \begin{bmatrix} 0 & -\omega_e & \frac{1}{C} & 0 \\ \omega_e & 0 & 0 & \frac{1}{C} \\ \frac{-1}{L} & 0 & \frac{-R}{L} & -\omega_e \\ 0 & \frac{-1}{L} & \omega_e & \frac{-R}{L} \end{bmatrix}\begin{bmatrix} v_{cq} \\ v_{cd} \\ i_q \\ i_d \end{bmatrix} + \begin{bmatrix} 0 \\ 0 \\ \frac{v_{tq}-e_{Bq}}{L} \\ \frac{v_{td}-e_{Bd}}{L} \end{bmatrix}$$

Per unit system is preferred in analysis and simulation. Hence, the above equation will be expressed in per unit system, where inductance and capacitance will be replaced by reactances. In the following equation, all variables and parameters are expressed in per unit, including ω_e:

$$\frac{d}{dt}\begin{bmatrix} v_{cq} \\ v_{cd} \\ i_q \\ i_d \end{bmatrix} = \omega_B \begin{bmatrix} 0 & -\omega_e & X_c & 0 \\ \omega_e & 0 & 0 & X_c \\ \frac{-1}{X_L} & 0 & -\frac{R_L}{X_L} & -\omega_e \\ 0 & \frac{-1}{X_L} & \omega_e & -\frac{R_L}{X_L} \end{bmatrix} \begin{bmatrix} v_{cq} \\ v_{cd} \\ i_q \\ i_d \end{bmatrix} + \omega_B \begin{bmatrix} 0 \\ 0 \\ \frac{v_{tq}-E_{Bq}}{X_L} \\ \frac{v_{td}-E_{Bd}}{X_L} \end{bmatrix}$$

where v_{cq} and v_{cd} are the quadrature and direct axis voltages across the capacitor, i_q and i_d are the quadrature and direct axis currents through the transmission line, v_{tq} and v_{td} are the quadrature and direct axis voltages of the terminal bus, E_{Bq} and E_{Bd} are the quadrature and direct axis voltages of the infinite bus, ω_B is the base speed (377 rad/s) and ω_e is the synchronous reference frame speed in per unit (1 pu). Note that network resonance at f_n will be observed as an oscillation mode with a complimentary frequency $f_s - f_n$ due to the synchronous reference frame. The state variables associated with the network are denoted by X_n and

$$X_n = [v_{cq}, v_{cd}, i_q, i_d]^T.$$

5.2 STATE-SPACE MODEL OF DFIG WIND ENERGY SYSTEM

5.2.1 Induction Generator Model

A seventh-order dynamic model [4] is used for the DFIG with rotor side converter. The model is derived from the voltage equations of an induction machine in a synchronous reference frame. The details of the model have been presented in Example 1 in Chapter 2 Section 2.4.1.

The details of building this model in Matlab/Simulink has also been covered in Chapter 2 as well as [4].

In addition to the induction machine model, DC-link dynamics and converter control dynamics will all be modeled for the SSR study system.

5.2.2 DC-Link Model

The dynamics of the capacitor in the DC link between the rotor and stator side converters are described by a first-order model (Fig. 5.2):

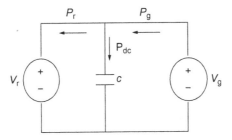

Figure 5.2 DC link, where $\overline{V}_r$ and $\overline{V}_g$ are the synthesized RSC and GSC voltage phasors. $\overline{V}_r = \frac{1}{\sqrt{2}}(v_{qr} - jv_{dr})$, $\overline{V}_g = \frac{1}{\sqrt{2}}(v_{qg} - jv_{dg})$. v_{qr} and v_{dr} are quadrature axis and direct axis RSC voltages. v_{qg} and v_{dg} are quadrature axis and direct axis GSC voltages.

$$Cv_{dc}\frac{dv_{dc}}{dt} = P_r - P_g \tag{5.1}$$

$$P_r = \frac{3}{2}(v_{qr}i_{qr} + v_{dr}i_{dr}) \tag{5.2}$$

$$P_g = \frac{3}{2}(v_{qg}i_{qg} + v_{dg}i_{dg}) \tag{5.3}$$

where P_r and P_g are the active power at RSC and GSC side, respectively.

5.2.3 Torsional Dynamics Model

A two-mass system popularly used to represent torsional dynamics [5] is given by

$$\frac{d}{dt}\begin{bmatrix} \Delta\omega_t \\ \Delta\omega_r \\ T_g \end{bmatrix} = \begin{bmatrix} \frac{-D_t-D_{tg}}{2H_t} & \frac{D_{tg}}{2H_t} & \frac{-1}{2H_t} \\ \frac{D_{tg}}{2H_g} & \frac{-D_g-D_{tg}}{2H_g} & \frac{1}{2H_g} \\ K_{tg}\omega_e & -K_{tg}\omega_e & 0 \end{bmatrix}\begin{bmatrix} \Delta\omega_t \\ \Delta\omega_r \\ T_g \end{bmatrix} + \begin{bmatrix} \frac{T_m}{2H_t} \\ \frac{-T_e}{2H_g} \\ 0 \end{bmatrix}$$

where ω_t and ω_r are the turbine and generator rotor speed, respectively; P_m and P_e are the mechanical power of the turbine and the electrical power of the generator, respectively; T_g is an internal torque of the model; H_t and H_g are the inertia constants of the turbine and the generator, respectively; D_t and D_g are the mechanical damping coefficients of the turbine and the generator, respectively; D_{tg} is the damping coefficient of the flexible coupling (shaft) between the two masses; K_{tg} is the shaft stiffness. The state variables associated with the torsional dynamics are denoted by X_t

$$X_t = [\Delta\omega_t, \Delta\omega_r, T_g]^T$$

The entire system model without consideration of converter controls (14th order) can be described by:

$$\dot{X} = f(X, U)$$

where $X = [X_n^T, X_g^T, v_{dc}, X_t^T]^T$ is assembled in Matlab/Simulink. Eigenvalue analysis and time domain simulations obtained with this model are described next.

5.2.4 DFIG Converter Controls
DFIG converter controls will be modeled in this study. Details about DFIG converter controls are presented in Chapter 3.

5.3 INTEGRATED SYSTEM MODEL

A key challenge in system model building is to integrate the network model and the machine model. The dynamic equations for the network have i_q, i_d, v_{cq} and v_{cd} as the state variables and have the terminal voltages of the DFIG v_{tq} and v_{td} as the input variables to the integrating block. There are two ways to take into consideration of the dynamics of the GSC current due to the inductor according to Padiyar [6].

For the Type-1 circuit (Fig. 5.3), the dynamics of the current through the inductor can be expressed as:

$$L\left(\frac{di_s}{dt} + \frac{di_e}{dt}\right) = v_{GSC} - v_t \tag{5.4}$$

For this type of circuit, the input voltage vector is expressed as the derivative of the state variables of both the generator and transmission line models. The dynamic model of the entire system becomes very complex.

The Type-2 circuit assumes that there is a shunt capacitor at the generator terminal (Fig. 5.3). This is true as the transmission line has shunt capacitance. For such kind of circuits, the dynamic model can be expressed as

$$L\frac{di_L}{dt} = v_{GSC} - v_t \tag{5.5}$$

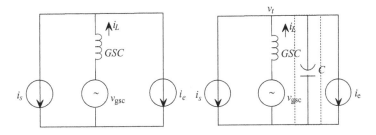

Figure 5.3 Equivalent circuit Type-1; Equivalent circuit Type-2.

$$C\frac{dv_t}{dt} = i_L - i_s - i_e \tag{5.6}$$

For the above system, let us examine the system response. The state-space model can be written as

$$\frac{d}{dt}\begin{bmatrix} i_L \\ v_t \end{bmatrix} = \begin{bmatrix} 0 & \frac{-1}{L} \\ \frac{1}{C} & 0 \end{bmatrix}\begin{bmatrix} i_L \\ v_t \end{bmatrix} + \begin{bmatrix} v_{GSC} \\ -i_s - i_e \end{bmatrix} \tag{5.7}$$

The system matrix has a pair of complex eigenvalues $\pm j1/\sqrt{LC}$ or $\pm j\omega_e\sqrt{X_C/X_L}$, where $\omega_e = 377$ rad/s. For an example study system, $X_L = 0.3\,pu$. The capacitance per phase to neutral for a 138 kV system is 0.186 MΩ/mile according to [7]. This value is used as the approximate capacitance per mile for the 161 kV system. Since the line length is 154 miles, the total capacitance is 28.64 MΩ. The impedance base for the 161 kV system with 100 MVA as the power base is 295 Ω. Hence $X_c = 110505$ pu. The eigenvalues are approximately $\pm j607\omega_e$.

The above analysis confirms that when the dynamics of the interfacing inductor are not ignored, a mode with a very high bandwidth will be introduced to the system. The bandwidth is way beyond the sub and super synchronous range. Hence, this dynamics can be ignored.

Further, we ignore the dynamics of the GSC inductor. This is reasonable since it is the network's LC resonance of major interest. That way, we can find out the terminal voltage phasor and qd variables based on the state variables of the differential equations.

The equivalent circuit with two current sources and one voltage source is shown in Fig. 5.4.

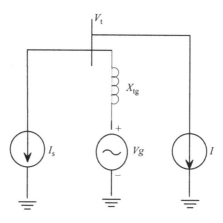

Figure 5.4 Equivalent circuit.

The relationship between the machine terminal voltage and the sources can be expressed as follows:

$$\overline{V}_t = \overline{V}_g - j(\overline{I}_s + \overline{I})X_{tg} \tag{5.8}$$

where X_{tg} is the reactance of the inductor that connects the GSC with the terminal machine bus, $\overline{V}_t = \frac{1}{\sqrt{2}}(v_{tq} - jv_{td})$, $\overline{V}_g = \frac{1}{\sqrt{2}}(v_{gq} - jv_{gd})$, and $\overline{I} = \frac{1}{\sqrt{2}}(I_q - jI_d)$.

The Matlab/Simulink block for the network dynamics is shown in Fig. 5.5, where the qd axis line currents and the qd axis capacitor voltages are the stator variables. The inputs of this block are the qd-axis machine terminal voltages and the outputs of this block are the qd-axis line currents.

The complete model that integrates the DFIG model and the network model is shown in Fig. 5.6. The outputs of the DFIG block are the stator currents and the GSC output voltages. Through the algebraic equation (5.8), the phasor of the terminal voltage $\overline{V}_t$ can be found using the outputs from both the machine and the network blocks. In turn, the qd-axis terminal voltages can also be found.

5.4 APPLICATION OF SSR ANALYSIS

5.4.1 Introduction

The phenomenon of SSR has been well studied in series-compensated systems with synchronous generators [8, 9]. SSR is divided into three

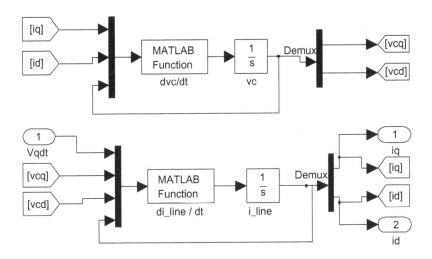

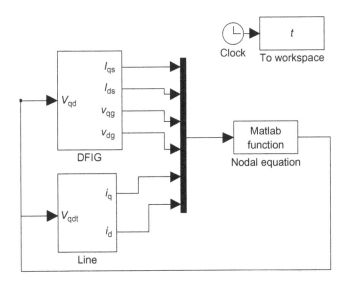

Figure 5.5 Matlab/Simulink block for the transmission line.

Figure 5.6 The complete Matlab/Simulink block.

groups: induction generator effect (IGE), torsional interaction (TI), and torsional amplification [9]. The first two denote phenomena related with steady state and the third denotes a transient phenomenon. While IGE involves interactions between the electrical network and the generator solely, torsional interactions include mechanical dynamics introduced by turbines.

SSR has been observed in Texas wind farms [10]. The phenomena can be analyzed using the state-space model developed in this chapter. In the following sections, we will first discuss the mechanism of SSR in DFIG. Then we will carry out small-signal analysis and time-domain simulations using the state-space model.

5.4.2 Analysis of SSR in a DFIG

A series-compensated network has a natural resonant frequency (f_n) given by $f_n = f_0 \sqrt{\frac{X_c}{X_L}}$ Hz (where f_0 is the synchronous frequency in Hz). At this subsynchronous frequency f_n, the slip s_1 is given by

$$s_1 = \frac{f_n - f_m}{f_n} \tag{5.9}$$

where f_m is the electric frequency corresponding to the rotating speed. Since f_n is usually less than f_m, $s_1 < 0$. From the steady state equivalent circuit for a DFIG (shown in Fig. 5.7), the equivalent rotor resistance at subsynchronous frequency is negative or $R_{r,eq} = R_r/s_1 < 0$. If the magnitude of this resistance exceeds the resistance of the armature plus network, the system has a negative resistance at the subsynchronous frequency. This can result in self-excitation leading to sustained or oscillatorily growing armature currents. This phenomenon is known as the IG effect [9].

Table 5.1 shows the frequency relationships in a DFIG. Due to series compensation, a subsynchronous component in voltages and currents with a frequency of f_n is introduced to the network and the stator circuit of a DFIG. The complementary frequency of this component in rotor current is ($f_s - f_n$). From Table 5.1, the stator currents have two components ($\overrightarrow{i_{s1}}$ and $\overrightarrow{i_{s2}}$) and the rotor currents also have two components ($\overrightarrow{i_{r1}}$ and $\overrightarrow{i_{r2}}$). $\overrightarrow{i_{s1}}$ is rotating at a nominal frequency f_s and $\overrightarrow{i_{s2}}$ is rotating at resonant frequency f_n. $\overrightarrow{i_{r1}}$ and $\overrightarrow{i_{r2}}$ are their corresponding rotor current components. These currents will have interactions to produce torque, which is shown in Table 5.2, where T_1 is due to the interaction of $\overrightarrow{i_{s1}}$ and $\overrightarrow{i_{r1}}$, T_2 is due to the interaction of $\overrightarrow{i_{s1}}$ and $\overrightarrow{i_{r2}}$, T_3 is due to the interaction of $\overrightarrow{i_{s2}}$ and $\overrightarrow{i_{r1}}$ and T_4 is due to the interaction of $\overrightarrow{i_{s2}}$ and $\overrightarrow{i_{r2}}$.

Thus, the torque will have a component at frequency of $f_s - f_n$ due to the network series compensation. Torsional interactions result when the frequency of the torsional mode f_{TI} is close or coincides with the complementary frequency $f_s - f_n$.

Table 5.1 Frequency Relationships in a DFIG		
Stator current space vector (stationary frame)	$\vec{I}_{s1}e^{j2\pi f_s t}$	$\vec{I}_{s2}e^{j2\pi f_n t}$
Rotating shaft speed	f_m	f_m
Rotor current space vector (stationary frame)	$\vec{I}_{r1}e^{j2\pi f_s t}$	$\vec{I}_{r2}e^{j2\pi f_n t}$
Rotor current space vector (rotor frame)	$\vec{I}_{r1}e^{j2\pi (f_s-f_m)t}$	$\vec{I}_{r2}e^{j2\pi (f_n-f_m)t}$

Table 5.2 Torque Components				
Torque comp.	T_1	T_2	T_3	T_4
Interaction	$\vec{i}_{s1}$ & $\vec{i}_{r1}$	$\vec{i}_{s1}$ & $\vec{i}_{r2}$	$\vec{i}_{s2}$ & $\vec{i}_{r1}$	$\vec{i}_{s2}$ & $\vec{i}_{r2}$
Frequency	0	$f_s - f_n$	$f_s - f_n$	0

The IGE is principally dependent on the slip s_1. The slip s_1 (from (5.9)) is influenced by wind speed and compensation level (from (5.9)). In addition, current control schemes which are used to obtain the rotor voltage injections can have impact on the equivalent rotor resistance $R_{r,eq}$ (from Fig. 5.7). The effect of these parametric variations on SSR phenomena is studied using the models developed in the following section.

5.4.3 Impact of Wind Speed and Compensation Level on SSR

The damping of the network resonant mode for different wind speeds are shown in Fig. 5.8. The results indicate that the damping of this mode improves with increasing wind speeds (and thus higher output powers). This is explained as follows. For a given wind speed, the optimal turbine rotational speed (f_m) is computed from the maximum power tracking scheme to maximize power extraction. When the wind speed decreases, this rotating speed f_m will decrease which in turn decreases the magnitude of slip s_1. Thus, s_1 is dependent on wind speed. Therefore, the equivalent rotor resistance $R_{r,eq}$ assumes a larger negative value and worsens the damping

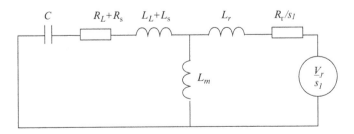

Figure 5.7 The equivalent circuit under subsynchronous frequency.

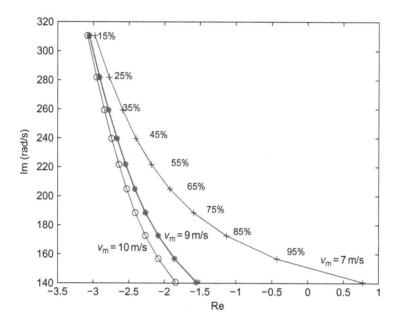

Figure 5.8 The network resonance mode at various compensation level for different wind speed. The embedded figure shows the equivalent rotor resistance versus the rotor rotating speed at 85% compensation level.

of this mode. The converse is true with increasing wind speeds. Figure 5.9 shows the relationship between the equivalent rotor resistance and the rotor speed. It is found that with higher wind speed, the higher f_m, hence $R_{r,eq}$ will be less negative. This feature is also noticeable from the dynamic response for two different wind speeds as shown in Fig. 5.11.

To focus on the IG effect, simulations are performed with the torsional system disabled. Figure 5.10 shows the dynamic response with varying compensation levels at a fixed wind speed (9 m/s). Figure 5.11 shows the dynamic response with varying wind speeds at a fixed compensation level (75%). The dynamic responses are initiated due to non-equilibrium initial conditions. It is also assumed that the rotor voltage injection is purely based on slip control. Observe from Fig. 5.10 that the higher the compensation level, the weaker the damping. From Fig. 5.11, the higher the wind speed, the better the SSR damping. This is in contrast to fixed speed systems where the damping of network mode worsens with increasing wind speeds. In fixed speed systems, the slip s_1 is independent of wind speed with f_m being very close to the nominal frequency f_s. Thus, the higher the wind power delivered to the system, the lower the resistance of the network as seen from the wind farm and hence the more unstable the mode as reported in [11].

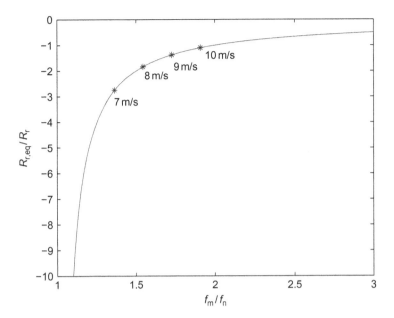

Figure 5.9 The equivalent rotor resistance versus the rotor rotating speed at 85% compensation level.

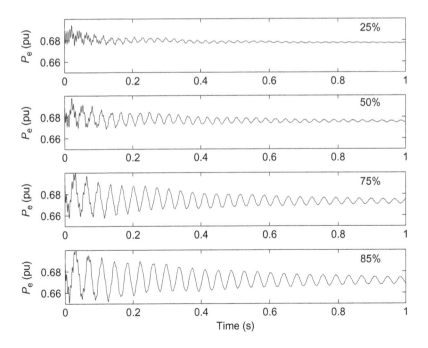

Figure 5.10 Dynamic response of P_e under different compensation level. Wind speeds: 9 m/s.

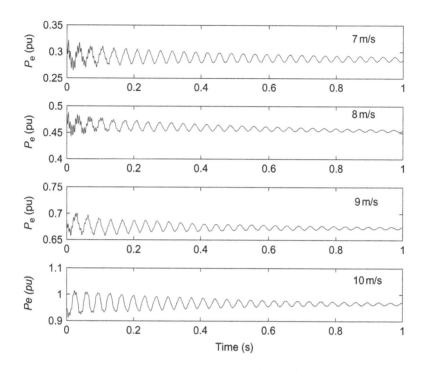

Figure 5.11 Dynamic response of P_e under different wind speeds. Compensation level: 75%.

With DFIG systems however, the damping of the network mode improves with increasing wind speeds as shown by the eignvalue results in Fig. 5.8 and simulations in Fig. 5.11. These findings are consistent with the observation that the dominant mode tends to become more unstable as the wind speed decreases, [12].

From Fig. 5.8, increasing levels of series compensation (at a fixed wind speed) decrease the damping of the SSR mode. The higher the compensation, the higher the natural resonant frequency f_n, and hence closer f_n will be to f_m. This will make the negative resistance have a larger magnitude and worsen the damping of the resonant modes [9].

5.4.4 Impact of DFIG Current Controllers on SSR

The rotor voltage injections in a DFIG are generated by current control loops in qd-axis as shown in Fig. 5.12 [4]. The reference current values are based on outer control loops for torque and voltage regulation [4]. Since current controllers are the focus of the investigation, for simplicity, the outer loops

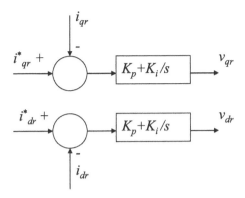

Figure 5.12 Current loop for rotor side converters.

are not modeled. Further, the integral unit gain K_i is set to zero. The effect of the proportional gain K_p will be investigated.

The effect of the proportional gains (K_p) in these control loops, on the damping of SSR mode is computed by eigenvalue analysis as shown in Table 5.3. The dynamic simulation results shown in Fig. 5.13 for P_e at different gains show that increasing gains have a detrimental impact on the damping. This is explained as follows. The effect of a rotor voltage injection $(V_r = kI_r)$ is equivalent to a resistance $R = \frac{-V_r/s}{I_r} = -\frac{k}{s}$ in the rotor circuit. A positive resistance will improve system stability. Since the slip s_1 under SSR is negative, then the gain k should be positive. However, in the DFIG current control schemes, negative feedback control loop is employed as shown in Fig. 5.12. Hence, the current control loops cause detrimental impact on the damping.

In order to keep the system from self-excitation, it is, therefore, necessary to keep the proportional gain of the current control loop in a range. Time-domain simulations are performed for various K_p in Fig. 5.13 corroborate the results in Table 5.3.

Table 5.3 Impact of DFIG Current Controllers on SSR at 75% Compensation Level with Wind Speed at 10 m/s			
K_p	Network mode	Hz	Damping (%)
0	$-2.3 \pm j180.6$	28.74	1.27
0.1	$9.8 \pm j185.6$	29.54	-5.21
0.5	$16.5 \pm j208.0$	33.10	-7.86

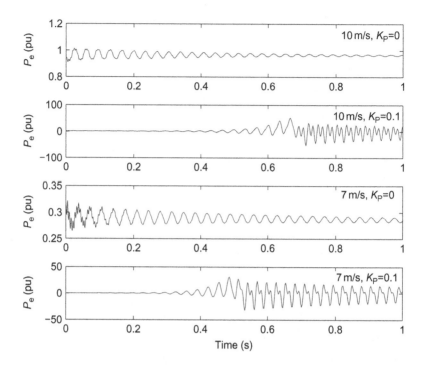

Figure 5.13 Dynamic response of P_e under different K_p. Compensation level: 75%

5.4.5 Results of TI Effect

The oscillatory mode of the torsional dynamics depends greatly on shaft stiffness K_{tg}. Wind turbine shafts have small stiffness constants [13] compared to the drive trains for steam, hydro and diesel generators. In the GE report [2], a 3.6-MW DFIG has a $K_{tg} = 3.6$ pu while in [11], this constant is set to 0.15 pu for a 100-MW drive train. The dependence of the torsional mode on shaft stiffness is tabulated in Table 5.4.

From Table 5.4, it is found that at normal range of K_{tg}, the frequency of torsional mode is less than 10 Hz. The complimentary frequency will be larger than 50 Hz in a 60 Hz AC network. It is unusual for the network resonant frequency to exceed 50 Hz, even at very high levels of compensation. Therefore torsional interactions are highly unlikely considering typical values of wind turbine shaft stiffness. For the sake of completeness and illustration, K_{tg} is chosen to be 99.67 pu (large) which results in a torsional oscillatory frequency at about 25 Hz and the complimentary frequency is 35 Hz. If by increasing the compensation level, the network resonant frequency gets close to 35 Hz, then TI is noticeable. In the following

Table 5.4 Torsional Modes for Various Stiffness Constant		
K_{tg}	Eigenvalue	$f_{TI}(Hz)$
0.15	-2.69 ± 31.74	5.05
1	-2.24 ± 33.70	5.36
2	-1.81 ± 36.26	5.77
3	-1.48 ± 38.93	6.20
10	-0.74 ± 56.19	8.94
50	-0.52 ± 115.28	18.35
100	-0.35 ± 162.45	25.85
150	-0.16 ± 193.79	30.84
200	-0.49 ± 225.42	35.87
250	-0.05 ± 252.1	40.12
Wind speed at 9m/s and compensation level at 75%.		

sections, impact of compensation level and wind speed is demonstrated by both eigenvalue analysis and time-domain simulations.

5.4.5.1 Impact of Compensation Level on Torsional Mode

In addition to the two-mass shaft system, the combined shaft and generator system has two modes of oscillations. One is the oscillation of the entire rotor against the power system; the other is the torsional mode. When the torsional mode frequency f_{TI} equals the complimentary of the network mode frequency $(f_s - f_n)$, the torsional interaction will occur [9]. When the series-compensation level is low, f_n is low. The complimentary frequency is high. If the series compensation increases, then f_n increases and the complimentary frequency decreases. Torsional interactions are noticeable when the frequency of the torsional mode is close or coincides with the complementary frequency $f_s - f_n$. Table 5.5 shows the torsional mode

Table 5.5 Comparison of the Torsional Mode and Network Mode at Various Series-Compensated Level with Multi-Mass Representation of the DFIG-Based Wind Turbine Rotor						
	Torsional Mode			Network Mode		
% Comp.	$\sigma \pm j\omega$	f_{TI} (Hz)	ξ (%)	$\sigma \pm j\omega$	$f_s - f_n$ (Hz)	ξ (%)
25	$-0.5 \pm j160.4$	25.53	0.31	$-2.9 \pm j270$	42.97	1.07
50	$-0.5 \pm j160.6$	25.56	0.31	$-2.6 \pm j221.3$	35.22	1.17
75	$-0.4 \pm j162.2$	25.81	0.25	$-2.3 \pm j178.8$	28.46	1.29
90	$5.2 \pm j158.9$	25.29	-3.27	$-7.6 \pm j158.4$	25.21	4.79

and the network mode with multi-mass representation of the DFIG-based wind turbine rotor. It is seen that f_{TI} is fairly insensitive for compensation levels between 25% to 75%. However, at 90% compensation, the mode is undamped (with frequency 25.29 Hz) and the frequency of network mode is 25.21 Hz. Therefore, there will be self excitation of SSR due to TI effect between the two modes. This is verified with the dynamic simulations shown in Fig. 5.16 where unstable oscillations are seen at the 90% compensation level. Additionally, the torsional mode becomes unstable (from 0.31% to -3.27%) while the damping of the network mode improves (from 1.07% to 4.79%). The locus of the torsional and network modes with various compensation levels are shown in Figs. 5.14 and 5.15, respectively.

5.4.5.2 Impact of Wind Speed on TI

The variation of network and torsional modes with wind speeds, at fixed compensation levels are shown in Table 5.6. In general, it can be noted that both the modes (in terms of frequency and damping) are fairly insensitive to changes in wind speed. This is explained as follows. The TI effect is due to the torsional dynamic oscillation at a frequency of f_{TI}. The network resonance frequency due to series compensation is f_n. When the wind speed

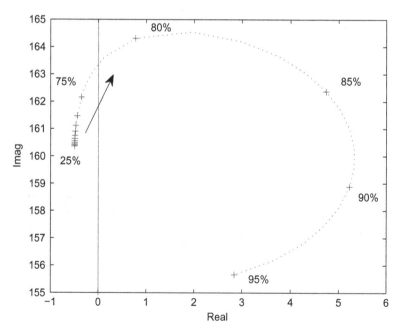

Figure 5.14 Torsional mode in the s-plane.

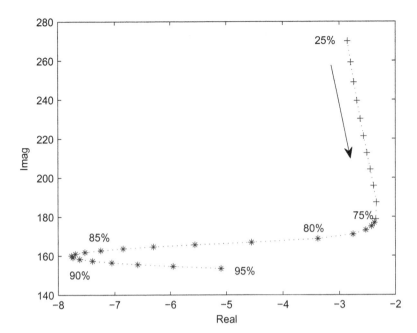

Figure 5.15 Network mode in the s-plane.

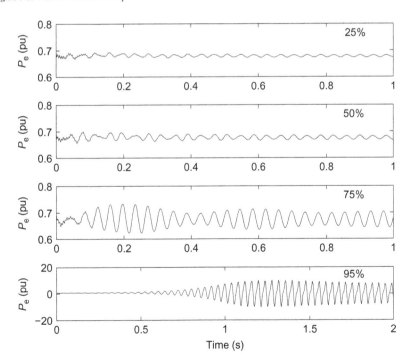

Figure 5.16 Electrical torque with various series-compensated in multi-mass DFIG-based wind turbine system. TI effects are demonstrated in the 90% case. Wind speed at 9 m/s.

Table 5.6 Torsional Mode and Network Mode at Various Wind Speeds

% Comp.	Wind Speed (m/s)	Torsional Mode			Network Mode		
		$\sigma \pm j\omega$	f_{TI} (Hz)	ξ (%)	$\sigma \pm j\omega$	$f_s - f_n$ (Hz)	ξ (%)
75%	7	$-0.4 \pm j164.7$	26.2	0.24	$-1.5 \pm j177.1$	28.19	0.85
	8	$-0.3 \pm j162.8$	25.91	0.18	$-2.1 \pm j178.3$	28.38	1.18
	9	$-0.4 \pm j162.2$	25.81	0.25	$-2.3 \pm j178.8$	28.46	1.29
	10	$-0.5 \pm j162.0$	25.78	0.31	$-2.5 \pm j179.0$	28.49	1.40
90%	7 m/s	$9.2 \pm j158.8$	25.27	-5.78	$-8.1 \pm j159.6$	25.40	5.1
	8	$6.2 \pm j159.0$	25.31	-3.9	$-8.1 \pm j158.5$	25.22	5.1
	9	$5.2 \pm j158.9$	25.29	-3.27	$-7.6 \pm j158.4$	25.21	4.79
	10	$4.7 \pm j159.0$	25.31	-2.95	$-7.4 \pm j158.3$	25.19	4.67

varies, the slip of the DFIG varies. This slip has no direct link to f_{TI} and f_n. Hence the wind speed has slight impact on SSR damping as shown in Table 5.6. Thus, the compensation level has a more significant impact on TI than the wind speed.

APPENDIX

Table 5.7 Parameters of a Single 2 MW DFIG and the Aggregated DFIG in Network System

Rated power	2 MW	100 MW
Rated voltage	690 V	690 V
X_{ls}	0.09231 pu	0.09231 pu
X_M	3.95279 pu	3.95279 pu
X_{lr}	0.09955 pu	0.09955 pu
R_s	0.00488 pu	0.00488 pu
R_r'	0.00549 pu	0.00549 pu
H	3.5 s	3.5 s
X_{tg}	0.3 pu (0.189 mH)	0.3 pu $\left(\frac{0.189}{5}\ \text{mH}\right)$
DC-link capacitor C	14,000 μF	50 $\times$ 14,000 μF
DC-link rated voltage	1200 V	1200 V

Table 5.8 parameters of the shaft system

H_1	0.9 s
H_2	4.29 s
D_1	0 pu
D_2	0 pu
D_{12}	1.5 pu
K_{12}	99.67 pu

REFERENCES

[1] EEE Subsynchronous Resonance Working Group, First benchmark model for computer simulation of subsynchronous resonance, IEEE Trans. Power Apparat. Syst. 96(5) (1977) 1565-1672.

[2] N.W. Miller, W.W. Price, J.J. Sanchez-Gasca, Dynamic modeling of GE 1.5 and 3.6 wind turbine-generators, GE-Power Systems Energy Consulting, General Electric International, Inc., Schenectady, NY, Oct. 2003.

[3] P. Krause, Analysis of Electric Machinery, McGraw-Hill, New York, 1986.

[4] Z. Miao, L. Fan, The art of modeling high-order induction generator in wind generation applications, Simulat. Model. Practice Theory 16(9) (2008) 1239-1253.

[5] F. Mei, B. Pal, Modal analysis of grid-connected doubly fed induction generators, IEEE Trans. Energy Convers. 22(3) (2007) 728-736.

[6] Padiyar, K. R. Analysis of subsynchronous resonance in power systems. Vol. 471. Springer Science & Business Media, 1998.

[7] Arthur R. Bergen and Vittal, Vijay. Power systems analysis. Prentice Hall, 1999.

[8] P. Kundur, Power System Stability and Control, McGraw-Hill Companies, New York, 1994.

[9] K.P. Padiya, Power System Dynamics Stability and Control. BS Publications, 2002.

[10] G.D. Irwin, A.K. Jindal, A.L. Isaacs, Sub-synchronous control interactions between type 3 wind turbines and series compensated ac transmission systems, in: 2011 IEEE Power and Energy Society General Meeting, IEEE, 2011, pp. 1-6.

[11] R.K. Varma, S. Auddy, Y. Semsedini, Mitigation of subsynchronous resonance in a series-compensated wind farm using FACTS controllers, IEEE Trans. Power Del. 23(3) (2008) 1645-1654.

[12] A. Ostadi, A. Yazdani, R. Varma, Modeling and stability analysis of a DFIG-based wind-power generator interfaced with a series-compensated line, IEEE Trans. Power Del. 24(3) (2009) 1504-1514.

[13] E.N. Hinrichsen, P.J. Nolan, Dynamics and stability of wind turbine generators, 101(8) (1982) 2640-2648.

Frequency-Domain Based DFIG Wind Energy Systems Modeling

6.1 INTRODUCTION OF IMPEDANCE MODEL AND ITS
 APPLICATION IN STABILITY AND RESONANCE DETECTION.... 95

6.2 IMPEDANCE MODELS OF DFIG IN VARIOUS
 REFERENCE FRAMES 96

6.2.1 RLC Circuit Impedance Model 97

6.2.2 Induction Machine Impedance Model 98

6.2.3 Converter Impedance Model 100

6.2.4 Overall Circuit 101

6.2.5 Comparison of Z_{sr} and Z_g 102

6.3 NEGATIVE-SEQUENCE DFIG IMPEDANCE MODEL 103

6.3.1 RLC Circuit 104

6.3.2 Induction Machine Impedance Model 104

6.3.3 Converter Impedance Model 105

6.3.4 Overall Circuit 106

6.4 INCLUSION OF DFIG INTO TORQUE-SPEED
 TRANSFER FUNCTIONS 106

6.4.1 Mechanical System 107

6.4.2 Electrical System 107

6.5 EXAMPLES.. 111

6.5.1 Example 1: SSR Detection in Series-Compensated
 Type-3 Wind Farms 111

6.5.1.1 Effect of Wind Speed 112

6.5.1.2 Effect of Compensation Degree..................................... 112

6.5.1.3 Effect of RSC PI Controller Gain 112

6.5.1.4 Validation Via Simulation 114

6.5.2 Example 2: Impact of Unbalance Operation on SSR.......... 119

Modeling and Analysis of Doubly Fed Induction Generator Wind Energy Systems.
http://dx.doi.org/10.1016/B978-0-12-802969-5.00006-8

6.5.3 Example 3: Torsional Interaction Investigation.................119

6.5.3.1 Frequency-Domain Analysis 119

6.5.3.2 Simulation Results..122

REFERENCES..127

In this chapter, DFIG's frequency-domain models at different reference frames (phase domain, *dq* reference frame) will be derived and explained. Then three examples on using the frequency-domain models for SSR analysis and torsional interaction analysis will be presented.

The chapter is organized in four sections. In Section 6.1, introduction of impedance model and its application in stability and resonance detection will be given. In Section 6.2, impedance models of DFIG in various reference frames, e.g., *abc* and *dq*, will be derived. In Section 6.3, negative-sequence impedance model is derived. In Section 6.4, inclusion of DFIG impedance model into a torque-speed transfer function will be presented. Section 6.5 presents three examples. The first example considers only balanced operating conditions and presents SSR detection using *abc*-domain impedance models. The second example considers unbalanced conditions and explains the impact of negative-sequence stator voltage on SSR using negative sequence impedance models. The third example uses torque-speed transfer function to explain DFIG's influence on torsional interactions.

6.1 INTRODUCTION OF IMPEDANCE MODEL AND ITS APPLICATION IN STABILITY AND RESONANCE DETECTION

Impedance-based models and Nyquist stability criterion provide insights on electric resonances. Such methods have been widely adopted in power electronics and system interaction studies [1, 2]. More recently, interactions between a grid-connected voltage source converter's current controllers and large network inductances are addressed in [3] using impedance models.

Impedance-based analysis has been applied in power electronic converter analysis in [1, 2]. For a system with a source impedance and load impedance as shown in Fig. 6.1, the current can be calculated as

$$I(s) = \frac{V(s)}{Z_s(s) + Z_l(s)} \tag{6.1}$$

$$= \frac{V(s)}{Z_l(s)} \cdot \frac{1}{1 + Z_s(s)/Z_l(s)} \tag{6.2}$$

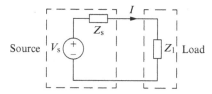

Figure 6.1 Small signal representation of a voltage source and a load [2].

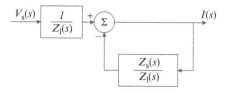

Figure 6.2 The converted feedback control system.

Assuming that the voltage source is stable and the load is stable when powered from an ideal voltage source, then for the system to be stable, the denominator $1 + Z_s(s)/Z_l(s)$ should have all zeros in the open left-half plane (LHP). Based on Nyquist stability criterion, if and only if the number of counter-clockwise encirclement around $(-1, 0)$ of Z_s/Z_l is equal to the number of the right-half plane (RHP) poles of Z_s/Z_l, the system will be stable. In cases when Z_s/Z_l has no RHP poles, the Nyquist map of Z_s/Z_l should not encircle $(-1, 0)$.

Instability happens when Z_s/Z_l encircles $(-1, 0)$. In addition, Bode plots can be used to analyze resonance stability. Bode plots of the source impedance and the load impedance can be placed together to identify the phase margin when the magnitudes of the two impedances are the same. This technique has been employed by Sun [2].

Impedance-based modeling technique converts a circuit analysis problem to a feedback control problem shown in Fig. 6.2. Analysis tools for linear control systems, e.g., Nyquist stability criteria, Bode plots, can then be deployed for stability and resonance analysis.

6.2 IMPEDANCE MODELS OF DFIG IN VARIOUS REFERENCE FRAMES

A study system consisting of a Type-3 DFIG-based wind generator with partial back-to-back voltage source converters and a series-compensated

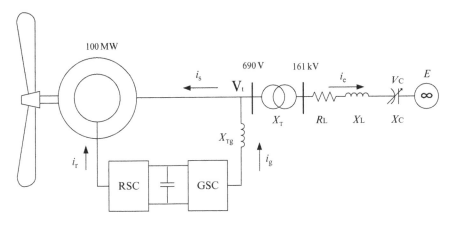

Figure 6.3 The study system. The rated power of the wind farm is 100 MVA. The nominal voltage of the wind farm terminal bus is 690 V and the nominal voltage of the network is 161 kV.

network is again shown in Fig. 6.3. A DFIG-based wind farm (100 MW from aggregation of 2 MW units) is connected to a 161 kV series-compensated line. The collective behavior of a group of wind turbines is represented by an equivalent lumped machine. X_{tg} represents the combination of the inductive filter at the GSC and a transformer.

The impedance models for the series-compensated network and the DFIG system will be developed in this section. For a three-phase system, impedance model can be expressed in either space vector or dq synchronous reference frame. A space vector voltage relates to the instantaneous three-phase voltages as follows:

$$\overrightarrow{V} = \frac{2}{3}\left(v_a + av_b + a^2v_c\right) \tag{6.3}$$

where $a = e^{j\frac{2\pi}{3}}$.

6.2.1 RLC Circuit Impedance Model

Assume that the series-compensated network has no mutual inductance and the three phases are symmetrical. For such a three-phase RLC circuit, the impedance model observed in space vector can be expressed as

$$Z_{net} = \frac{\overrightarrow{V}}{\overrightarrow{I}} = \frac{v_a}{i_a} = R + sL + \frac{1}{sC} \tag{6.4}$$

6.2.2 Induction Machine Impedance Model

To find out the impedance model of a DFIG, we start to find out the relationship between the stator voltage and the stator current. Essentially, it is just to find the equivalent Thevenin impedance and the equivalent Thevenin voltage for the induction machine circuit in space vector (shown in Fig. 6.4).

A few steps can lead us to the stator voltage expression in terms of stator current and rotor voltage solely.

Observe the dynamic model in space vector:

$$\vec{v_s} = R_s \vec{i_s} + \frac{d\vec{\psi_s}}{dt} \tag{6.5}$$

$$\vec{v_r} = R_r \vec{i_r} + \frac{d}{dt}\vec{\psi_r} - j\omega_m \vec{\psi_r} \tag{6.6}$$

$$\vec{\psi_s} = L_s \vec{i_s} + L_m \vec{i_r} \tag{6.7}$$

$$\vec{\psi_r} = L_m \vec{i_s} + L_r \vec{i_r}. \tag{6.8}$$

We will first find the expression of $\vec{i_r}$ by $\vec{i_s}$ and $\vec{v_r}$ from the rotor voltage and rotor flux relationship, shown in (6.9). Then in the stator voltage and stator flux relationship, the rotor current will be replaced by the following expression:

$$\vec{i_r} = \frac{\vec{v_r} - (s - j\omega_m)L_m \vec{i_s}}{R_r + (s - j\omega_m)L_r} \tag{6.9}$$

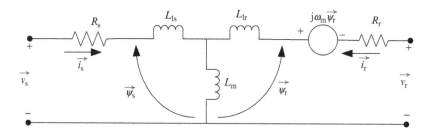

Figure 6.4 Induction machine circuit model in space vector.

The stator voltage can be expressed as follows:

$$\vec{v_s}(s) = \underbrace{\left[R_s + s\left(L_s - \frac{L_m^2[s - j\omega_m)]}{R_r + [s - j\omega_m)L_r]} \right) \right]}_{Z} \vec{i}_s + \frac{sL_m}{R_r + (s - j\omega_m)L_r} \vec{v}_r$$

(6.10)

Note a space vector based on the stationary frame is related with a complex vector based on the qd reference frame (rotating speed is ω) is as follows:

$$\vec{f} = \bar{F}_{qds}e^{j\omega t}.$$

(6.11)

In the frequency domain, their relation becomes

$$\vec{f}(s) = \bar{F}_{qds}(s - j\omega) \qquad \vec{f}(s + j\omega) = \bar{F}_{qds}(s).$$

(6.12)

Therefore, in complex vector, the stator voltage is expressed by replacing s in (6.10) with $s + j\omega$.

$$\bar{V}_s = \underbrace{\left[R_s + (s + j\omega)\left(L_s - \frac{L_m^2[s + j(\omega - \omega_m)]}{R_r + [s + j(\omega - \omega_m)L_r]} \right) \right]}_{Z} \bar{I}_s$$

$$+ \frac{(s + j\omega)L_m}{R_r + [s + j(\omega - \omega_m)]L_r} \bar{V}_r$$

If L_m is very large compared to the other parameters, then the above equation can be simplified as

$$Z_s(s) = R_s + s(L_{ls} + L_{lr}) + \frac{s}{s - j\omega_m}R_r.$$

(6.13)

The above derivation corroborates with the following analysis on slip. Slip is related to the rotating speed ω_m and the stator frequency ω. For the per-phase induction machine circuit shown in Fig. 6.5, the rotating speed can be assumed to be constant since mechanical dynamics is much slower than the electric dynamics. Slip can be expressed as $1 - \omega_m/\omega$. In Laplace domain, slip can be expressed as

$$\mathrm{slip}(s) = \frac{s - j\omega_m}{s}.$$

(6.14)

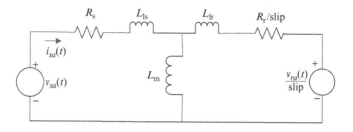

Figure 6.5 The per-phase circuit of an induction machine.

Hence, the impedance of the DFIG seen from its terminal can be represented by

$$Z_{\text{DFIG}}(s) = R_r/\text{slip}(s) + R_s + (L_{ls} + L_{lr})s. \tag{6.15}$$

The above derivation also proves that the impedance model can be developed using the per-phase circuit of a DFIG. The simplified DFIG impedance in (6.15) ignoring RSC and GSC is adopted in [4].

6.2.3 Converter Impedance Model

Cascaded control loops are used in converters in wind generators. The control loops consist of the fast inner current control loops and the slow outer power/voltage control loops. The current control loops usually have bandwidths at or greater than 100 Hz while the outer control loops usually have bandwidths less than several Hz. For SSR studies, the dynamics of interest are much faster than the outer control loops. Thus, the outer control is considered to be constant and will not be included in the impedance model.

For the vector current control scheme in Fig. 6.6, the qd-axis voltage and current relationship is

$$v_q = (i_q^* - i_q)H_i(s) - K_d i_d \tag{6.16}$$

$$v_d = (i_d^* - i_d)H_i(s) + K_d i_q \tag{6.17}$$

This leads to the expression in complex vector

$$\bar{V}_{qds} = \bar{I}_{qds}^* H(s) - (H(s) - jK_d)\bar{I}_{qds}. \tag{6.18}$$

To view the current controlled converter from space vector, then

$$\overrightarrow{V} = \overrightarrow{I}^* H(s - j\omega) - (H(s - j\omega) - jK_d)\overrightarrow{I} \tag{6.19}$$

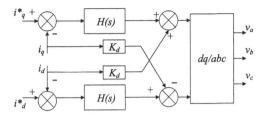

Figure 6.6 The converter current control loops.

For the GSC at positive sequence scenarios, it can be represented by a voltage source $\overrightarrow{I}_g^* H_g(s - j\omega)$ behind an impedance Z_{GSC}, where $Z_{GSC} = H_g(s - j\omega) - jK_{dg}$.

For the RSC at positive sequence scenarios, it can also be represented by a voltage source $\overrightarrow{I}_r^* H_r(s - j\omega)$ behind an impedance Z_{RSC} where $Z_{RSC} = H_r(s - j\omega) - jK_{dr}$.

6.2.4 Overall Circuit

The overall circuit is now shown in Fig. 6.7. The circuit consists of two impedances: the network impedance Z_{net} and the DFIG impedance Z_{DFIG}.

When the gains of the RSC controller K_p K_i and the feed forward gain K_{dr} are assumed to be zero, the RSC output voltage will no longer vary even there exists error in measured currents and the reference currents. The RSC can be viewed as a constant voltage source in this case.

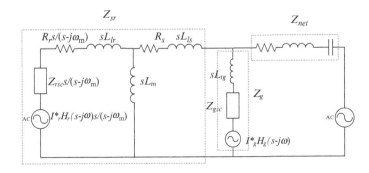

Figure 6.7 The overall circuit model.

6.2.5 Comparison of Z_{sr} and Z_g

The DFIG impedance consists of two parallel components: the stator and rotor (including RSC) Z_{sr} and the GSC branch including the GSC and the filter: $Z_g = Z_{GSC} + sL_{tg}$. Bode plots of the RLC network impedances, Z_{sr} and Z_g at 95% rotating speed (or 57 Hz) and 50% compensation level are plotted in Fig. 6.8. Z_{sr} is plotted for two scenarios: with and without current PI controllers. When the PI controller gain is not zero, Z_{sr} has two peaks: one at 57 Hz and the other at 60 Hz. When the gain is zero, Z_{sr} however has just one peak at 57 Hz. For Z_g, the resonant frequency is 60 Hz. The network impedance has a resonant frequency at 37.5 Hz, which correspond to the 50% compensation level. The network resonant frequency is expressed as:

$$f_n = 60\sqrt{\frac{X_c}{X_{line} + X_T}} \tag{6.20}$$

where X_c is the series capacitor reactance, X_{line} and X_T are the line reactance and the transformer reactance.

It is observed that at the interested SSR frequency region (<37.5 Hz), the magnitude of Z_g is much larger than Z_{sr}. Therefore, Z_{sr} is dominant. For the SSR analysis, the impact of Z_g will be ignored ($Z_{DFIG} = Z_{sr}$).

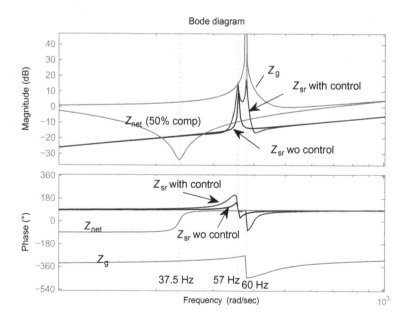

Figure 6.8 Bode plots of Z_g, Z_{net}, Z_{sr}.

6.3 NEGATIVE-SEQUENCE DFIG IMPEDANCE MODEL

When the system has both positive and negative sequence components, and let $\bar{V}_p$ and $\bar{V}_n$ be the positive and negative sequence components in $\bar{V}_a$ where $\bar{V}_a$ is the dynamic phasor of the phase a voltage at the fundamental frequency.

The phasors are related to the instantaneous variables as follows:

$$
\begin{cases}
v_a(t) = (\bar{V}_p + \bar{V}_n)e^{j\omega t} + (\bar{V}_p^* + \bar{V}_n^*)e^{-j\omega t} \\
v_b(t) = (a^2\bar{V}_p + a\bar{V}_n)e^{j\omega t} + (a\bar{V}_p^* + a^2\bar{V}_n^*)e^{-j\omega t} \\
v_c(t) = (a\bar{V}_p + a^2\bar{V}_n)e^{j\omega t} + (a^2\bar{V}_p^* + a\bar{V}_n^*)e^{-j\omega t}
\end{cases}
\tag{6.21}
$$

where $a = e^{j2\pi/3}$.

The space vector of the terminal voltage is defined as

$$
\overrightarrow{V}(t) = \left(\frac{2}{3}\right)\left(v_a(t) + av_b(t) + a^2 v_c(t)\right).
\tag{6.22}
$$

The space vector can also be expressed in terms of positive and negative sequence phasors

$$
\overrightarrow{V}(t) = 2\bar{V}_p e^{j\omega t} + 2\bar{V}_n^* e^{-j\omega t}.
\tag{6.23}
$$

In Laplace domain, the voltage and current space vectors can be written as

$$
\begin{cases}
\overrightarrow{V}(s) = 2\bar{V}_p(s - j\omega) + 2\bar{V}_n^*(s + j\omega) \\
\overrightarrow{I}(s) = 2\bar{I}_p(s - j\omega) + 2\bar{I}_n^*(s + j\omega)
\end{cases}
\tag{6.24}
$$

When there is only positive sequence or negative sequence, then the impedance observed based on space vectors should have the following relationship with the impedance observed based on the complex vectors or dynamic phasor.

$$
\begin{cases}
\text{Positive: } Z_p = \dfrac{\overrightarrow{V}(s)}{\overrightarrow{I}(s)} = \dfrac{\bar{V}_p(s-j\omega)}{\bar{I}_p(s-j\omega)} \\[3mm]
\text{Negative: } Z_n = \dfrac{\overrightarrow{V}(s)}{\overrightarrow{I}(s)} = \dfrac{\bar{V}_n^*(s+j\omega)}{\bar{I}_n^*(s+j\omega)}
\end{cases}
\tag{6.25}
$$

Since the expression of $\frac{\bar{V}_p}{\bar{I}_p}$ and $\frac{\bar{V}_n}{\bar{I}_n}$ are the same, it can be found that

$$Z_p = Z_n^*. \tag{6.26}$$

6.3.1 RLC Circuit

For a three-phase RLC circuit, the impedance model observed in space vector can be expressed as:

$$Z_{\text{line,p}} = Z_{\text{line,n}} = R + sL + \frac{1}{sC} \tag{6.27}$$

The above expression also fits (6.26). Since there is no imaginary part for the positive sequence expression, the negative sequence and the positive sequence expressions are the same.

6.3.2 Induction Machine Impedance Model

We already have the relationship between the stator voltage and stator current, rotor voltage in dq reference frame:

$$\bar{V}_s = \underbrace{\left[R_s + (s+j\omega)\left(L_s - \frac{L_m^2[s+j(\omega-\omega_m)]}{R_r + [s+j(\omega-\omega_m)L_r]} \right) \right]}_{Z} \bar{I}_s$$

$$+ \frac{(s+j\omega)L_m}{R_r + [s+j(\omega-\omega_m)]L_r} \bar{V}_r. \tag{6.28}$$

Therefore, a Thevenin equivalent circuit for the induction machine can be developed. Notice that the impedance $Z(s)$ is based on the dq-reference frame. Since the RLC circuit is expressed in the abc reference frame, it will be convenient that there is also an impedance model based on the abc reference frame.

In space vector the stator voltage can be expressed as follows by replacing s by $s - j\omega$ for positive sequence.

Positive sequence:

$$\overrightarrow{V}(s) = \bar{V}_s(s-j\omega)$$

$$= \underbrace{\left[R_s + s\left(L_s - \frac{L_m^2(s-j\omega_m)}{R_r + (s-j\omega_m)L_r]} \right) \right]}_{Z} \overrightarrow{I}_s + \frac{sL_m}{R_r + (s-j\omega_m)L_r} \overrightarrow{V}_r$$

$$\tag{6.29}$$

Therefore, the positive sequence impedance model is

$$Z_{DFIG,p} = R_s + s \left(L_{ls} + L_m - \frac{L_m^2}{\dfrac{R_r}{s - j\omega_m} + L_{lr} + L_m} \right)$$

$$= R_s + sL_{ls} + sL_m \frac{\dfrac{R_r}{s - j\omega_m} + L_{lr}}{\dfrac{R_r}{s - j\omega_m} + L_{lr} + L_m} \tag{6.30}$$

The negative sequence impedance model is

$$Z_{DFIG,n} = Z_{DFIG,p}^* = R_s + s \left(L_{ls} + L_m - \frac{L_m^2}{\dfrac{R_r}{s + j\omega_m} + L_{lr} + L_m} \right)$$

$$= R_s + sL_{ls} + sL_m \frac{\dfrac{R_r}{s + j\omega_m} + L_{lr}}{\dfrac{R_r}{s + j\omega_m} + L_{lr} + L_m} \tag{6.31}$$

6.3.3 Converter Impedance Model

Cascaded control loops are used in converters in wind generators. The control loops consist of the fast inner current control loops and the slow outer power/voltage control loops. The current control loops usually have bandwidths at or greater than 100 Hz while the outer control loops usually have bandwidths less than several Hz. For SSR studies, the study dynamics is considered faster than the outer control loops. Thus, the outer control is considered to be constant and will not be included in the impedance model.

For the vector current control scheme in Fig. 6.6, the qd-axis voltage and current relationship is

$$v_q = (i_q^* - i_q)H_i(s) - K_d i_d \tag{6.32}$$

$$v_d = (i_d^* - i_d)H_i(s) + K_d i_q \tag{6.33}$$

This leads to the expression in complex vector:

$$\bar{V}_{qd} = \bar{I}_{qd}^* H_i(s) - (H_i(s) - jK_d)\bar{I}_{qd}. \tag{6.34}$$

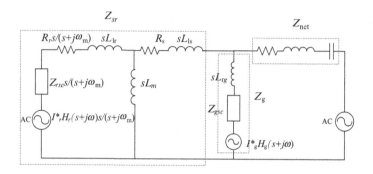

Figure 6.9 The overall circuit model for negative sequence.

To view the current controlled converter from space vector, then:

$$\begin{cases} \text{Positive: } \overrightarrow{V} = \overrightarrow{I}^*H_i(s - j\omega) - (H_i(s - j\omega) - jK_d)\overrightarrow{I} \\ \text{Negative: } \overrightarrow{V} = \overrightarrow{I}^*H_i(s + j\omega) - (H_i(s + j\omega) + jK_d)\overrightarrow{I} \end{cases} \quad (6.35)$$

For the GSC converter at positive sequence scenarios, it can be represented by a voltage source $\overrightarrow{I}_g^*H_g(s - j\omega)$ behind an impedance Z_{GSC} where $Z_{GSC} = H_g(s - j\omega) - jK_{dg}$. At negative sequence scenarios, the GSC converter can be represented by a voltage source $\overrightarrow{I}_g^*H_g(s + j\omega)$ behind an impedance Z_{GSC} where $Z_{GSC} = H_g(s + j\omega) + jK_{dg}$.

For the RSC converter at positive sequence scenarios, it can also be represented by a voltage source $\overrightarrow{I}_r^*H_r(s - j\omega)$ behind an impedance Z_{RSC} where $Z_{RSC} = H_r(s - j\omega) - jK_{dr}$. The negative sequence impedance of the RSC is the conjugate of Z_{RSC} and can be expressed as $H_r(s + j\omega) + jK_{dr}$.

6.3.4 Overall Circuit
The overall circuit for negative sequence is now shown in Fig. 6.9.

6.4 INCLUSION OF DFIG INTO TORQUE-SPEED TRANSFER FUNCTIONS

To investigate the impact of negative sequence components on torsional interaction, transfer function of the electromagnetic torque (T_e) versus the rotating speed (ω_m) is developed in this section. The interactions of torque and speed relationship is represented in Fig. 6.10 [5], where $G_e(s)$ represents the relationship of T_e versus ω_m through electrical system and $G_m(s)$

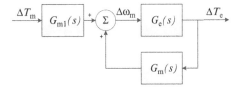

Figure 6.10 Torque and rotating speed relationship.

represents the relationship of ω_m and T_e through mechanical system. Note the positive feedback is due to the motor convention used in this chapter. When operated as a generator, T_e is a negative number.

Therefore the electromagnetic torque and speed transfer function is:

$$G_{cl}(s) = \frac{G_e(s)}{1 - G_m(s)G_e(s)}. \qquad (6.36)$$

6.4.1 Mechanical System

A two-mass system [6, 7] is used to the represent torsional dynamics given by

$$
\begin{bmatrix} \Delta\dot{\omega}_t \\ \Delta\dot{\omega}_m \\ \dot{T}_{12} \end{bmatrix} =
\begin{bmatrix}
\frac{-D_t-D_{tg}}{2H_t} & \frac{D_{tg}}{2H_t} & \frac{-1}{2H_t} \\
\frac{D_{tg}}{2H_g} & \frac{-D_g-D_{tg}}{2H_g} & \frac{1}{2H_g} \\
K_{tg}\omega_e & -K_{tg}\omega_e & 0
\end{bmatrix}
\begin{bmatrix} \Delta\omega_t \\ \Delta\omega_m \\ T_{12} \end{bmatrix} +
\begin{bmatrix} \frac{T_m}{2H_t} \\ \frac{-T_e}{2H_g} \\ 0 \end{bmatrix}.
$$

where ω_t and ω_r are the turbine and generator rotor speed, respectively; P_m and P_e are the mechanical power of the turbine and the electrical power of the generator, respectively; T_{12} is an internal torque of the model; H_t and H_g are the inertia constants of the turbine and the generator, respectively; D_t and D_g are the mechanical damping coefficients of the turbine and the generator, respectively; D_{tg} is the damping coefficient of the flexible coupling (shaft) between the two masses; K_{tg} is the shaft stiffness.

From the two-mass mechanical system model, the transfer functions $G_{m1} = \frac{\Delta\omega_m}{\Delta T_m}$ and $G_m = \frac{\Delta\omega_m}{\Delta T_e}$ can be found.

6.4.2 Electrical System

Torque-speed transfer function has been used in [8] to examine torsional interactions. The voltage and current relationship for an induction machine

(2.53) in the dq reference frame are linearized. These equations are presented again here.

$$
\begin{bmatrix} v_{qs} \\ v_{ds} \\ v_{0s} \\ v_{qr} \\ v_{dr} \\ v_{0r} \end{bmatrix} =
$$

$$
\begin{bmatrix}
R_s + \frac{P}{\omega_b}X_s & \frac{\omega_s}{\omega_b}X_s & 0 & \frac{P}{\omega_b}X_m & \frac{\omega_s}{\omega_b}X_m & 0 \\
-\frac{\omega_s}{\omega_b}X_s & R_s + \frac{P}{\omega_b}X_s & 0 & -\frac{\omega_s}{\omega_b}X_m & \frac{P}{\omega_b}X_m & 0 \\
0 & 0 & R_s + \frac{P}{\omega_b}X_{ls} & 0 & 0 & 0 \\
-\frac{P}{\omega_b}X_m & \frac{\omega_s-\omega_m}{\omega_b}X_m & 0 & R_r + \frac{P}{\omega_b}X_r & \frac{\omega_s-\omega_m}{\omega_b}X_r & 0 \\
-\frac{\omega_s-\omega_m}{\omega_b}X_m & \frac{P}{\omega_b}X_m & 0 & -\frac{\omega_s-\omega_m}{\omega_b}X_r & R_r + \frac{P}{\omega_b}X_r & 0 \\
0 & 0 & 0 & 0 & 0 & R_r + \frac{P}{\omega_b}X_{lr}
\end{bmatrix}
$$

$$
\begin{bmatrix} i_{ds} \\ i_{qs} \\ i_{0s} \\ i_{qr} \\ i_{dr} \\ i_{0r} \end{bmatrix}
\tag{6.37}
$$

where $\omega_b = 377$ rad/s and the rest variables are in per unit.

The resulting linearized differential equations are as follows:

$$
\frac{d}{dt}\Delta \mathbf{i}_{sr} = A_1 \Delta \mathbf{i}_{sr} + A_2 \Delta \omega_m + B_1 \Delta \mathbf{v}_s + B_2 \Delta \mathbf{v}_r
\tag{6.38}
$$

where $\mathbf{i}_{sr} = [i_{qs}, i_{ds}, i_{qr}, i_{dr}]^T$, $\mathbf{v}_s = [v_{qs}, v_{ds}]^T$ and $\mathbf{v}_r = [v_{qr}, v_{dr}]^T$,

$$
A_1 = -\left(\frac{1}{\omega_b}\begin{bmatrix} X_s & 0 & X_m & 0 \\ 0 & X_s & 0 & X_m \\ X_m & 0 & X_r & 0 \\ 0 & X_m & 0 & X_r \end{bmatrix} \right)^{-1}
\begin{bmatrix} R_s & X_s & 0 & X_m \\ -X_s & R_s & -X_m & 0 \\ 0 & sX_m & R_r & sX_r \\ -sX_m & 0 & -sX_r & R_r \end{bmatrix},
\tag{6.39}
$$

$$
A_2 = -\left(\frac{1}{\omega_b}\begin{bmatrix} X_s & 0 & X_m & 0 \\ 0 & X_s & 0 & X_m \\ X_m & 0 & X_r & 0 \\ 0 & X_m & 0 & X_r \end{bmatrix} \right)^{-1}
\begin{bmatrix} 0 & 0 & 0 & 0 \\ 0 & 0 & 0 & 0 \\ 0 & -X_m & 0 & -X_r \\ X_m & 0 & X_r & 0 \end{bmatrix},
\tag{6.40}
$$

$$B_1 = \frac{\omega_b}{\sigma X_s X_r} \begin{bmatrix} X_r & 0 \\ 0 & X_r \\ -X_m & 0 \\ 0 & -X_m \end{bmatrix}$$ (6.41)

where $\sigma = 1 - \frac{X_m^2}{X_s X_r}$,

$$B_2 = \frac{\omega_b}{\sigma X_s X_r} \begin{bmatrix} -X_m & 0 \\ 0 & -X_m \\ X_s & 0 \\ 0 & X_s \end{bmatrix}.$$ (6.42)

In Laplace domain, the above equation becomes:

$$(sI - A_1)\Delta i_{sr} = A_2 \Delta \omega_m + B_1 \Delta v_s + B_2 \Delta v_r$$ (6.43)

Consider the current control loops in the RSC, the RSC voltage can be expressed as:

$$\Delta v_r = H_{RSC}(s) \Delta i_r^* - Z_{RSC} \Delta i_r$$ (6.44)

If we assume that the converter outer control loops can be ignored, then the reference signal i_r^* are constants and its deviation is zero.

Substituting $\Delta v_r'$ in (6.43) by (6.44) leads to:

$$\Delta i_{sr} = (sI - A_1 + B_2 \begin{bmatrix} 0 & Z_{RSC} \end{bmatrix})^{-1} (A_2 \Delta \omega_m + B_1 \Delta v_s)$$

$$= G_{vi} \Delta v_s + G_{\omega i} \Delta \omega_m.$$ (6.45)

It is obvious to find the expression of Δi_s in terms of the rotating speed and the stator voltage:

$$\Delta i_s = G_{vis} \Delta v_s + G_{\omega is} \Delta \omega_m.$$ (6.46)

The expressions can be developed for both positive and negative sequences. For positive sequence, the DQ reference frame (dq^+) is rotating counter clockwise at nominal frequency ω. While for negative sequence, the DQ reference frame is rotating clockwise at the nominal frequency ω. The major differences in expressions are A_1 and Z_{RSC}. For positive sequence, since the vector control is based on the same reference frame, $Z_{RSC}^+ = H(s) = K_p + K_i/s$. However, for negative sequence, since the vector control is based on dq^+ while the model is based on dq^-, the impedance of the RSC should be converted to qd^-. Hence $Z_{RSC} = H(s-j2\omega) = K_p + K_i/(s-j2\omega)$.

$$\begin{cases} \Delta i_{s,p}^+ = G_{vis}^+ \Delta v_{s,p} + G_{\omega is,p}^+ \Delta \omega_m \\ \Delta i_{s,n}^- = G_{vis}^- \Delta v_{s,n} + G_{\omega is,n}^- \Delta \omega_m \end{cases} \tag{6.47}$$

where the superscripts "+" and "−" represent the $dq+$ and $dq-$ rotating reference frames.

The electromagnetic torque of an induction machine can be expressed by the stator and the rotor currents.

$$T_e = \frac{X_M}{2}(i_{qs}i_{dr} - i_{ds}i_{qr}). \tag{6.48}$$

When there is only positive sequence, T_e is a DC variable. When there is only negative sequence, T_e is also a DC variable. To evaluate the impact of positive and negative sequence components on the DC component of T_e, the following definitions are made:

$$\begin{cases} T_{e,dc1} = \frac{X_M}{2}(i_{qs,p}^+ i_{dr,p}^+ - i_{ds,p}^+ i_{qr,p}^+) \\ T_{e,dc2} = \frac{X_M}{2}(i_{qs,n}^- i_{dr,n}^- - i_{ds,n}^- i_{qr,n}^-) \end{cases} \tag{6.49}$$

Note that the positive sequence components in the $dq+$ frame are DC at steady-state while the negative-sequence components in the $dq-$ frame are DC at steady-state.

The linearized expressions are as follows:

$$\Delta T_{e,dc1} = \frac{X_M}{2}\begin{bmatrix} i_{dr,p}^+ & -i_{qr,p}^+ & -i_{ds,p}^+ & i_{qs,p}^+ \end{bmatrix} \Delta i_{sr,p}^+ = G_{vt}^+ \Delta v_{s,p}^+ + G_{\omega t}^+ \Delta \omega_m \tag{6.50}$$

$$\Delta T_{e,dc2} = \frac{X_M}{2}\begin{bmatrix} i_{dr,n}^- & -i_{qr,n}^- & -i_{ds,n}^- & i_{qs,n}^- \end{bmatrix} \Delta i_{sr,n}^- = G_{vt}^- \Delta v_{s,n}^- + G_{\omega t}^- \Delta \omega_m \tag{6.51}$$

In addition, the stator voltage can be expressed by the stator current through the network equation if the current through the GSC is ignored:

$$\begin{cases} \Delta v_{s,p}^+ = -Z_{line}^+ \Delta i_{s,p}^+ \\ \Delta v_{s,n}^- = -Z_{line}^- \Delta i_{s,n}^- \end{cases} \tag{6.52}$$

where

$$\begin{cases} Z^+_{\text{line}} = \begin{bmatrix} R + sL + \dfrac{s}{(s^2 + \omega^2)C} & \omega L - \dfrac{\omega}{(s^2 + \omega^2)C} \\ -\omega L + \dfrac{\omega}{(s^2 + \omega^2)C} & R + sL + \dfrac{s}{(s^2 + \omega^2)C} \end{bmatrix} \\[4em] Z^-_{\text{line}} = \begin{bmatrix} R + sL + \dfrac{s}{(s^2 + \omega^2)C} & -\omega L + \dfrac{\omega}{(s^2 + \omega^2)C} \\ \omega L - \dfrac{\omega}{(s^2 + \omega^2)C} & R + sL + \dfrac{s}{(s^2 + \omega^2)C} \end{bmatrix} \end{cases} \tag{6.53}$$

According to (6.51), (6.47) and (6.52), the torque versus the speed transfer functions $G_e(s)$ can be found:

$$G^+_e(s) = -G^+_{vt}(I + Z^+_{\text{line}} G^+_{vis})^{-1} Z^+_{\text{line}} G^+_{wis} + G^+_{wt}$$

$$G^-_e(s) = -G^-_{vt}(I + Z^-_{\text{line}} G^-_{vis})^{-1} Z^-_{\text{line}} G^-_{wis} + G^-_{wt}$$

So far, we have given the torque/speed diagram's transfer functions block by block. The model will be useful to investigate torsional interactions. The derived model has included the effect of DFIG and its rotor converter control.

6.5 EXAMPLES

6.5.1 Example 1: SSR Detection in Series-Compensated Type-3 Wind Farms

In this example, Nyquist criterion based stability analysis will be carried out to study the effect on SSR due to wind speed, compensation degree and RSC PI controller gain. The expression of Z_{sr} can be found from the overall circuit shown in Fig. 6.7. The open-loop transfer function to be studied is Z_{net}/Z_{sr} and its expression is as follows:

$$\frac{Z_{\text{net}}}{Z_{sr}} = \cfrac{r + sL + \frac{1}{sC}}{R_s + sL_{ls} + \cfrac{sL_m \left(sL_{lr} + \dfrac{s}{s - j\omega_m}(R_r + K_p + \dfrac{K_i}{s - j\omega}) - jK_{dr}\right)}{sL_m + sL_{lr} + \dfrac{s}{s - j\omega_m}(R_r + K_p + \dfrac{K_i}{s - j\omega}) - jK_{dr}}} \tag{6.54}$$

6.5.1.1 Effect of Wind Speed

Using the same study system and same parameters as in [9], two impedances are derived and the Nyquist map of Z_{net}/Z_{sr} is plotted in Fig. 6.11. The gains of the RSC controllers are set to zeros. The compensation level is 75%. It is shown that when the rotating speed at 0.75 pu, the Nyquist map will encircle $(-1, 0)$ which indicates instability. When the rotating speed is at 0.85 and 0.95 pu, the Nyquist maps will not encircule $(-1, 0)$ and the system is marginally stable. The phase margins are 2° and 5° for the 0.85 and 0.95 pu rotating speed. The points when the Nyquist maps intersect with the unit circuit have frequencies of about 40 Hz. This indicates the resonance frequency. It also corresponds to the LC resonant frequency at 75% compensation level.

The Nyquist map demonstrates the effect of wind speed on SSR stability since low wind speed corresponds to low rotating speed. Hence a Type-3 wind generator is more prone to SSR when wind speed is lower.

6.5.1.2 Effect of Compensation Degree

The Bode plots of the network impedance Z_{net} at different compensation degrees and the DFIG impedance Z_{sr} are plotted in Fig. 6.12. The rotating speed is assumed to be 0.75 pu. It is observed that a higher compensation degree results in a higher frequency at which $Z_{sr} = Z_{net}$: 251 rad/s (40 Hz) at 75% versus 206 rad/s (32.8 Hz) at 50%. It is also observed that the phase angle of Z_{sr} at that frequency is 107° at 75% compensation level compared to 95° at 50% compensation level. Therefore, Z_{net}/Z_{sr} will have a less phase margin or become unstable when the compensation level is higher.

The Nyquist maps of the network Z_{net} and the DFIG impedance Z_{sr} at different compensation degrees are plotted in Fig. 6.13. The points where the Nyquist maps intersects the unit circle are at 33 Hz for 50% compensation level (phase margin (5°) and 40 Hz for 75% compensation level. The Nyquist maps indicate that (-1,0) is encircled at 75% compensation level. Hence, a higher compensation degree results in a less stable system.

6.5.1.3 Effect of RSC PI Controller Gain

In this section, the effect of the RSC PI controller gain is examined. The time constant of the PI controller is fixed at 0.02 s. The wind speed is selected as 9 m/s and the corresponding rotating speed is 0.95 pu. The compensation level is 50%. Two cases are examined.

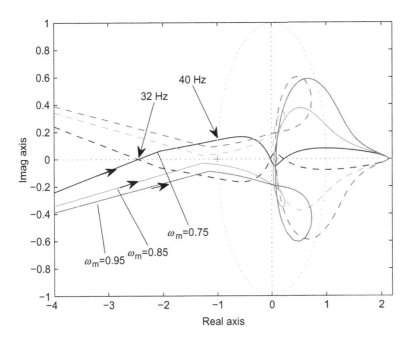

Figure 6.11 Nyquist map for different rotating speeds. Series compensation level: 75%. Gains of the current controllers are zeros.

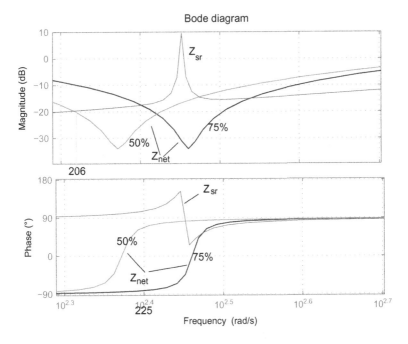

Figure 6.12 Bode plots for Z_{net} at different compensation degree, Z_{sr} at Rotating speed 0.75 pu.

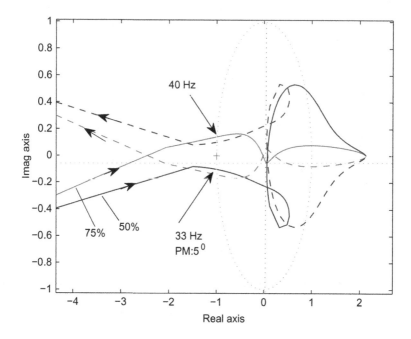

Figure 6.13 Nyquist map for different compensation degree. Rotating speed: 0.75 pu.

- Case 1 $K_p = 0$
- Case 2 $K_p = 0.01$

The Nyquist plots are shown in Figs. 6.14 and 6.15. It is observed that when the gain is larger or equal 0.01, the Nyquist maps encircle $(-1, 0)$ clockwise, which indicates instability. The point when the curve traverses the real axis is at 35 Hz. When the gain is zero, $(-1, 0)$ is not encircled and the phase margin is 11° at 33 Hz.

Bode plots of the impedances Z_{net} and Z_{sr} are shown in Fig. 6.16. It is shown that a higher current controller gain increases the magnitude of Z_{sr} and leads to a lower frequency at $Z_{net} = Z_{sr}$. The phase angle of the impedances at unit gain of Z_{net}/Z_{sr} are recorded in Table 6.1.

It is found that a higher gain leads to less phase margin or instability.

6.5.1.4 Validation Via Simulation
The nonlinear model presented in Chapter 5 has been built in Matlab/Simulink. The impact of wind speed and the RSC current loop control on

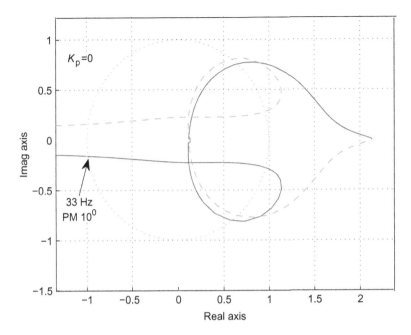

Figure 6.14 Nyquist map when $K_p = 0$. Compensation level 50%. Rotating speed: 0.95 pu.

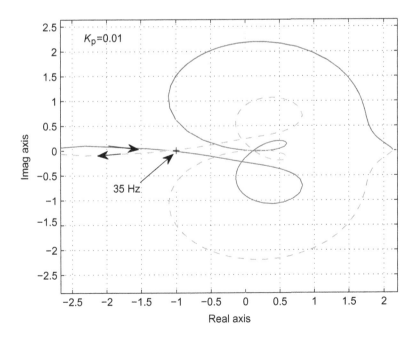

Figure 6.15 Nyquist map when $K_p = 0.01$. Compensation level 50%. Rotating speed: 0.95 pu.

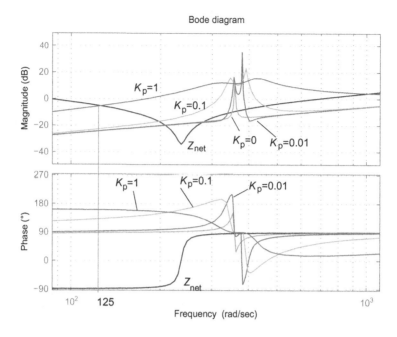

Figure 6.16 Bode plots of Z_{sr} at different current controller gain.

Table 6.1 Phase Angles of the Impedances and Stability Detection				
	$K_p = 1$	$K_p = 0.1$	$K_p = 0.01$	$K_p = 0$
Res. freq. (Hz)	20	32	35	33
$\angle Z_{net}$	$-88°$	$-81.3°$	$-77°$	$-77°$
$\angle Z_{sr}$	161°	149°	99.4°	91.4°
$\angle Z_{net}/Z_{sr}$	111°	130°	$-178°$	$-168.4°$
Stability	N	N	Marginal	stable

SSR have also been discussed. It is observed that at lower wind speed, with high compensation level, the system is subject to SSR. In this simulation study, controller interaction between the RSC's current control and the electrical network will be verified. In [10], such phenomenon is classified as subsynchronous controller interaction (SSCI). The case study scenarios will be setup with wind speed at 9 m/s and the rotating speed at 95% of the nominal. The compensation level at 50%. The RSC control gain will be varied to show the impact of controller interaction.

- Case 1 $K_p = 0.0$
- Case 2 $K_p = 0.01$

The system is initially operated at 25% compensation level. At $t = 1$ s, additional capacitors are switched in to make the compensation level reach 50%. Figures 6.17–6.19 show the dynamic responses of the system with thick lines refer to Case 1 while the thin lines refer to Case 2. The dynamic responses of the electromagnetic torque, the rotating speed of the wind turbine and the terminal voltage of the DFIG wind farm are shown Fig. 6.17. It is found that there is a resonance at about 25 Hz in the voltage magnitude and torque. For Case 1 the resonance can be damped out in the torque and the terminal voltage after 1 s. However, for case 2, the resonance cannot be damped out after 1 s.

Figure 6.18 shows the dynamic responses of the DFIG wind farm output power P_e and Q_e, the voltage across the capacitors V_c and the current through the line I_e. The resonance is damped out in Case 1 while persists in Case 2.

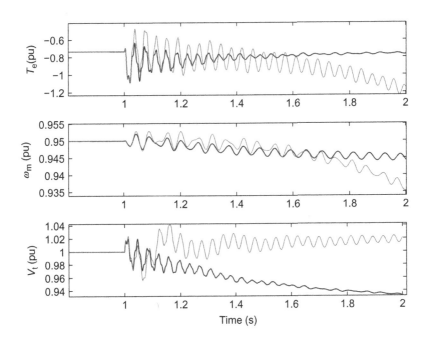

Figure 6.17 Dynamic response of T_e, ω_m and V_t. Thick lines: Case 1. Thin lines: Case 2.

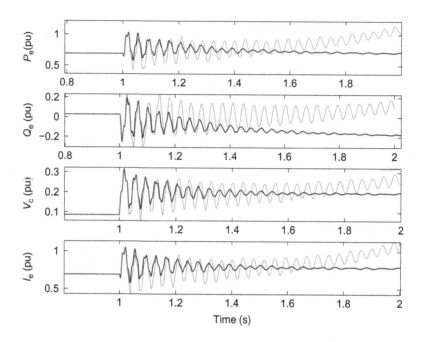

Figure 6.18 Dynamic response of P_e, Q_e, V_c and I_e. Thick lines: Case 1. Thin lines: Case 2.

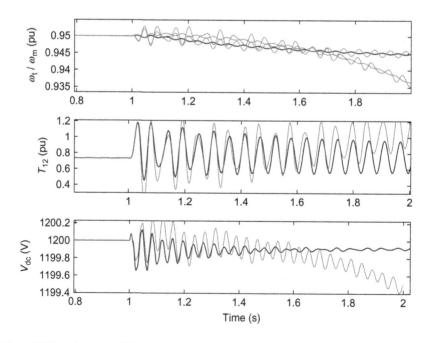

Figure 6.19 Dynamic response of T_e, ω_m/ω_t and V_{dc}. Thick lines: Case 1. Thin lines: Case 2.

Figure 6.19 shows the dynamic responses of the rotating speeds of two masses ω_t and ω_m, the torque T_{12} between the two masses of the wind turbine and the DC-link voltage V_{dc}. The resonance is more obvious in Case 2. In addition, power/torque imbalance is observed that the rotating speed and the DC-link voltage keep decreasing.

According to the Nyquist/Bode analysis in Section III, the resonance is 35 Hz at 50% compensation level and 0.95 pu rotating speed. This resonance reflected in power, torque and voltage/current magnitude will have a frequency of 25 Hz. The analysis in Fig. 6.15 also shows marginal stability or instability when the gain K_p is 0.01. Simulation results demonstrate the effect of current controller gain.

6.5.2 Example 2: Impact of Unbalance Operation on SSR

The bode plots of the negative sequence impedances are shown in Fig. 6.20. Two network impedances are presented: one is at 25% compensation level while the other is at 70% compensation level. Two DFIG impedances are presented: one is at 75% nominal speed while the other is at 95% nominal speed. From the Bode plots, the two Bode plots almost match each other. Therefore wind speed has negligible impact on the DFIG impedance. The network impedance magnitudes and the DFIG impedance magnitudes meet at 150 rad/s (24 Hz, 25% compensation level) and 250 rad/s (40 Hz, 75% compensation level). There are phase margin for Z_{line}/Z_{DFIG}. Hence the negative sequence impedance will not cause SSR instability.

A comparison of the positive and negative sequence impedances of the DFIG is shown in Fig. 6.21. From this figure, it is obvious that the positive sequence impedance of the DFIG could be greater than 90° at the gain cross frequencies. This indicates that it is the positive sequence impedance that could cause SSR instability.

6.5.3 Example 3: Torsional Interaction Investigation
6.5.3.1 Frequency-Domain Analysis

Bode plots of the torque/speed transfer function $G_{cl}(s)$ will be plotted. The operating conditions are: wind speed 9 m/s, compensation level (25%, 50%, and 75%). Two scenarios will be examined: with positive sequence components only and with negative sequence components only. In the first case, $G_{cl}(s) = \dfrac{G_e^+(s)}{1 - G_m(s)G_e^+(s)}$ and in the second case, $G_{cl}(s) = \dfrac{G_e^-(s)}{1 - G_m(s)G_e^-(s)}$.

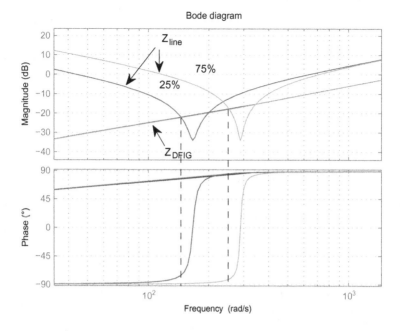

Figure 6.20 Bode plots of negative sequence impedances.

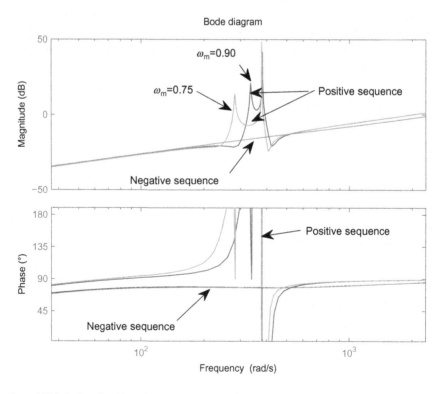

Figure 6.21 Bode plots of positive and negative sequence impedances.

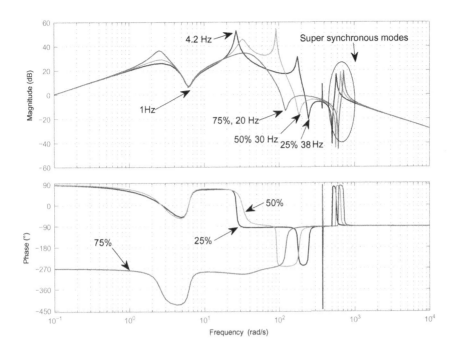

Figure 6.22 Bode plots of torque/speed relationship considering positive sequence components only. Wind speed 9 m/s.

Figure 6.22 presents Bode plots of torque/speed relationship considering positive sequence components only. Three scenarios with different compensation levels are presented. The Bode plots indicate that the system has several oscillation modes, including a mode related to mechanical dynamics with less than 1 Hz, a mode related to the electromechanical interaction at 4-6 Hz, the subsynchronous resonance (SSR) mode ($\sim$ 38Hz at 25%, $\sim$ 30Hz at 50% and $\sim$ 20Hz at 75%) and the supersynchronous mode. Note that the frequency of the SSR mode in torque has the complementary frequency of the SSR frequency in voltages and currents. For example, the SSR frequency of the electric system at 9 m/s at 25% compensation level is 24 Hz. The resonance frequency can be found in Figure 6.20 the phase domain impedance Bode plot. In Fig. 6.22, the SSR frequency of 9 m/s 25% case is also around 36 Hz.

Figure 6.23 presents Bode plots of torque/speed relationship considering negative sequence components only. These Bode plots are obtained based on the linearized model where an initial negative sequence operating condition is assumed. The Bode plots in Fig. 6.22 are also shown here in dotted lines

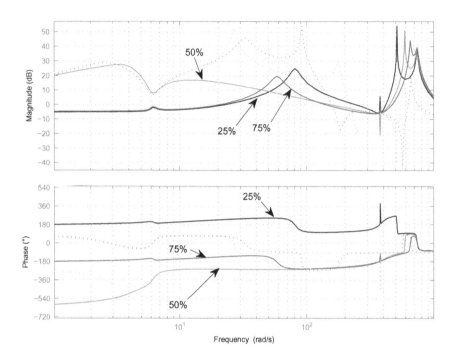

Figure 6.23 Bode plots of torque/speed relationship considering only negative sequence components. Wind speed 9 m/s. Dotted lines correspond to positive sequence components.

as comparison. It can be seen that magnitude wise, the negative sequence component's effect in the range of 160 Hz is insignificant compared to the positive sequence component.

6.5.3.2 Simulation Results

In this section, time-domain simulation is carried out in Matlab/Simulink using nonlinear dq based models developed in Chapter 5. Two types of disturbances are applied to the test system where a Type-3 wind farm interconnected to a series-compensated transmission line). The parameters of the test system can be found in the Appendix of Chapter 5.

The wind speed is chosen to be 9 m/s and the system compensation level is 25%. Three cases are studied.

- In Case 1, at $t = 2$s, a positive sequence voltage drop (0.1 pu) is applied at the system voltage.
- In Case 2, at $t = 2$s, a negative sequence disturbance (0.1 pu) is applied to the system voltage.

- In Case 3, at $t = 2s$, a negative sequence disturbance (0.5 pu) is applied to the system voltage.

The simulation model is built in dq synchronous reference frame. In Case 1, a step response is applied at the q-axis system voltage. While in Case 2 and Case 3, a step response is applied at the q-axis of system voltage of the dq^- reference frame. It is then transformed into dq^+ reference frame as shown in Fig. 6.24.

The system dynamic responses will be presented. The simulation results are presented in the following figures. Figure 6.25 presents the machine rotating speed dynamic responses. It can be observed that there are SSR mode at 36 Hz and a 4.5 Hz oscillation in the balanced disturbance case. When unbalanced disturbance is applied, the 120 Hz oscillation is sustained as long as the unbalanced disturbance is presented. In addition, a 4.5 Hz oscillation is also obvious. The SSR oscillation distorts the 120 Hz oscillation as well. At 2.5 s, the distortion is no longer obvious, which indicates the die-out of the SSR mode.

Figure 6.26 presents the torsional speed (ω_t) and the intermediate torque T_{12} in the two-mass turbine. A much lower oscillation mode (<1 Hz) is now presented. This mode corresponds to the mechanical mode indicated in the Bode plots. The 4.5 Hz oscillation can be observed obviously. The comparison of three cases indicate that a 0.5 pu unbalanced disturbance has similar effect of 0.1 pu balanced disturbance. Regarding the effect on damping of these two modes, unbalanced or balanced disturbance does not have any noticed difference.

As a comparison, another scenario when the series compensation level is 50% is studied and the simulation results are presented in Figs. 6.27 and 6.28. Observing Case 1 in Fig. 6.27 can find that the SSR mode is

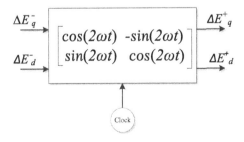

Figure 6.24 dq^- to dq^+.

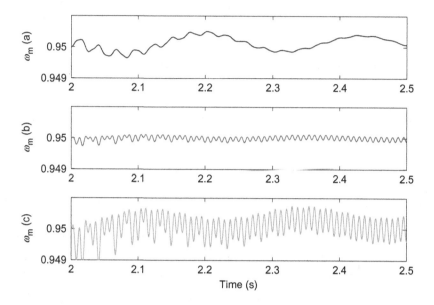

Figure 6.25 Rotating speed when subject to a system voltage disturbance. a) Case 1; b) Case 2; c) Case 3.

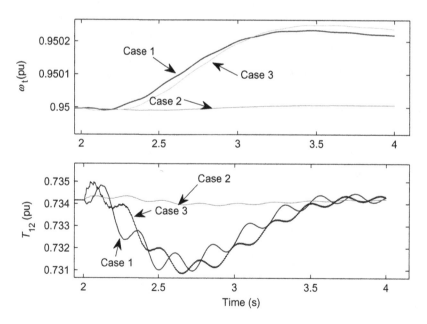

Figure 6.26 Torsional speed and intermediate torque.

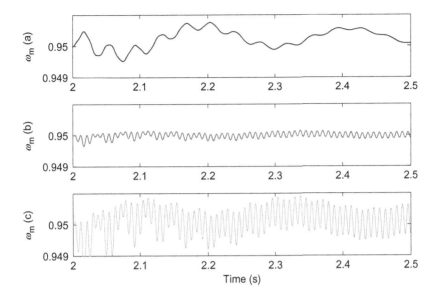

Figure 6.27 Rotating speed when subject to a system voltage disturbance. Compensation level 50%. (a) Case 1; (b) Case 2; (c) Case 3.

now with a frequency of 27 Hz. This observation corroborates with the analysis by Bode plots in Section 6.5.3.1 (e.g. Fig. 6.22) that when series compensation level increases, the SSR mode frequency decreases in torque. Comparison of the dynamic response of T_{12} in Fig. 6.28 indicates that the 0.5- and 5-Hz oscillations are all presented in balanced and unbalanced cases.

Figure 6.29 present the simulation results of electric power and electromagnetic torque. In both cases, 36 Hz SSR mode can be observed in the balanced case and unbalanced case. Though unbalanced case has sustained 120-Hz oscillation.

The above simulation results demonstrate that except an addition of 120 Hz oscillation, unbalanced disturbance does not worsen the other oscillation modes. It is also understandable that the simulation case is a small disturbance case, therefore the system characteristic will be determined by the operating point which is balanced operation.

Comparison cases also indicate that a 5 times larger unbalanced disturbance could have comparable impact as a balanced disturbance on the system if the 120 Hz oscillation is excluded.

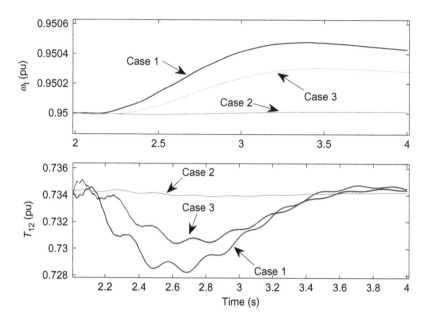

Figure 6.28 Torsional speed and intermediate torque. Compensation level 50%.

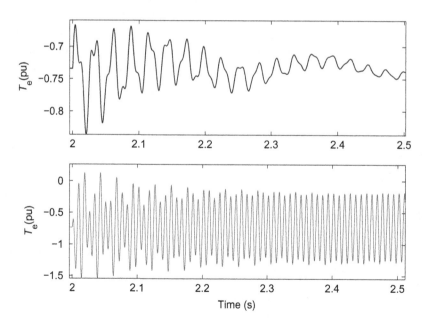

Figure 6.29 Dynamic response of electromagnetic torque of DFIG. Compensation level 25%. (a) Case 1; (b) Case 3.

REFERENCES

[1] J. Sun, Small-signal methods for AC distributed power systems a review, IEEE Trans. Power Electron. 24(11) (2009) 2545-2554.

[2] J. Sun, Impedance-based stability criterion for grid-connected inverters, IEEE Trans. Power Electron. 26(11) (2011) 3075-3078.

[3] M. Cespedes, J. Sun, Modeling and mitigation of harmonic resonance between wind turbines and the grid, in: IEEE Energy Conversion Congress and Exposition (ECCE), Sep. 2011.

[4] Lingling Fan; Zhixin Miao, "Nyquist-Stability-Criterion-Based SSR Explanation for Type-3 Wind Generators," Energy Conversion, IEEE Transactions on, vol. 27, no. 3, pp. 807, 809, Sept. 2012.

[5] A. Tabesh, R. Iravani, Frequency-response analysis of torsional dynamics, IEEE Trans. Power Syst. 19(3) (2004) 1430-1437.

[6] W. Qiao, W. Zhou, R.G. Harley, Wind speed estimation based sensorless output maximization control for a wind turbine driving a DFIG, 23(3) (2008) 1156-1169.

[7] P. Kundur, Power System Stability and Control, McGraw-Hill Companies, New York, 1994.

[8] L. Harnefors, Analysis of subsynchronous torsional interaction with power electronic converters, IEEE Trans. Power Syst. 22(1) (2007) 305-313.

[9] L. Fan, R. Kavasseri, Z. Miao, C. Zhu, Modeling of DFIG-based wind farms for ssr analysis, IEEE Trans. Power Del. 25(4) (2010) 2073-2082.

[10] G. Irwin, A. Jindal, A. Isaacs, Sub-synchronous control interactions between type 3 wind turbines and series compensated AC transmission systems, in: IEEE Power & Energy General Meeting 2011, Jul. 2011.

CHAPTER 7

Multi-Machine Modeling and Inter-Area Oscillation Damping

7.1 STEADY-STATE CALCULATION FOR A DFIG 129

7.2 INTERCONNECTION OF DFIG MODEL IN POWER SYSTEMS 132

7.3 APPLICATION OF THE MODEL: INTERAREA-OSCILLATION
DAMPING THROUGH DFIG WIND TURBINES 133

7.3.1 Configuration of the Test System 134

7.3.2 Plant Model 134

7.3.3 Damping Control of DFIG.. 136

7.3.4 Validation Through Time-Domain Simulation 140

APPENDIX 144

REFERENCES.. 145

In this chapter, dq-reference frame-based modeling of a power system with a DFIG is explained. The application of such model is demonstrated through a control design example for inter-area oscillation damping. The control design includes a few steps: (i) identification of the input and the output to obtain the plant model, (ii) model reduction, (iii) root-locus based design, and (iv) verification by nonlinear simulation.

The first emphasis of this chapter is the technique of interfacing a DFIG with a conventional synchronous generator dominated power system. Dynamic modeling of synchronous generators will not be discussed in this chapter. Interested readers can refer classic textbooks [1, 2] on synchronous generator modeling and multi-machine power system modeling. To have a small-signal or a linear model for control design, we need to find the equilibrium point or the current operating condition. For conventional power systems, the operating condition can be found by conducting load flow analysis. Further, based on the steady-state synchronous generator circuit models, internal variables of a generator can be found, e.g., qd-axis stator currents and voltages, rotor angle, etc. For a system integrated with a DFIG, finding the operating condition needs load flow calculation as well as DFIG

Modeling and Analysis of Doubly Fed Induction Generator Wind Energy Systems.
http://dx.doi.org/10.1016/B978-0-12-802969-5.00007-X

initialization. The DFIG initialization can be conducted through circuit analysis of an equivalent DFIG circuit model. The initialization part is addressed in Section 7.1. The interfacing technique is addressed in Section 7.2.

The second emphasis of this chapter is the application of the dynamic model. The application presented in this chapter is small-signal analysis. Further, control design based on the linearized model is presented. The derivation of the small-signal model is a standard procedure as follows.

Given $\dot{x} = f(x, u)$, the linearized model at certain operation condition (x_0, u_0) is

$$\Delta \dot{x} = \left. \frac{\partial f}{\partial x} \right|_{x_0, u_0} \Delta x + \left. \frac{\partial f}{\partial u} \right|_{x_0, u_0} \Delta u \qquad (7.1)$$

Given a continuous dynamic model, the corresponding linearized model can be obtained by MATLAB in one command "linmod." Therefore, derivation will not be discussed. Instead, throughout this book, we assume that the readers can use tools to develop a linearized model once the nonlinear continuous dynamic model is given.

7.1 STEADY-STATE CALCULATION FOR A DFIG

Initial condition calculation of state variables is required to find an equilibrium point. In dynamic simulation, if we initialize the state variables with values from a stable equilibrium point, the simulation should appear as flat run when there is no dynamic event imposed. Initial condition is also important for small-signal analysis since linearization is based on this operating condition. The underlying assumption of small-signal analysis and initialization is that the state variables are constant at steady state. For a dynamic model based on *abc* instantaneous currents and voltages, linearization is not possible since at steady-state the waveforms of currents and voltages are periodic. In the *dq*-reference frame, when the system is balanced, the steady-state currents and voltages are constants. Therefore, *dq*-based models are suitable for small-signal analysis.

For a synchronous generator in a power system, after its output real and reactive powers as well as terminal voltage magnitude are given, using the well-developed steady-state generator model or phasor diagrams, we can find internal variables such as initial rotor position, internal voltage magnitude, stator current phasor, flux linkages, etc. Interested readers can consult a classic textbook such as [3, Chapter 6].

In power flow problems, a generator bus can be either treated as a PV bus where the active power and voltage magnitude are scheduled, or a PQ bus where both active power and reactive power outputs are scheduled. After a power flow problem is solved, for every bus in the system, its voltage phasor, active power, and reactive power injection become known.

Here for a DFIG terminal bus, we will assume that the terminal voltage magnitude (V_s) is known. Its total active power (P_{DFIG}) and reactive power (Q_{DFIG}) to the grid are also known. Our task is to find the internal variables for a DFIG, including stator current phasor $\bar{I}_s$, rotor current phasor $\bar{I}_r$, and the rotor voltage phasor $\bar{V}_r$. Most importantly, we need to determine the speed of the wind generator (ω_m) or the slip ($s = 1 - \omega_m$). We assume the wind turbine is working at its maximum power extracting mode.

In total there are seven unknown real variables to be found. We need to define seven equations for these seven unknown variables.

By observing the equivalent circuit of a DFIG machine in Fig. 7.1, we can find two equations in the complex domain or four equations in terms of real variables by applying KVL. These are

$$\bar{V}_s = -(R_s + j(X_{ls} + X_m))\bar{I}_s + jX_m\bar{I}_r \tag{7.2}$$

$$\bar{V}_r = (R_r + js(X_{lr} + X_m))\bar{I}_r - jsX_m\bar{I}_s. \tag{7.3}$$

where $s = 1 - \omega_m$.

We would also like to express the total active power and total reactive power by the unknown variables. GSC control can regulate its output reactive power. Therefore this reactive power can be assumed as known. Therefore, the reactive power from the stator Q_s can be assumed as known. Here are the two additional equations:

$$P_{DFIG} = P_s - P_r = V_sI_s \cos(\theta_s - \phi_s) - V_rI_r \cos(\theta_r - \phi_r) \tag{7.4}$$

$$Q_s = V_sI_s \sin(\theta_s - \phi_s) \tag{7.5}$$

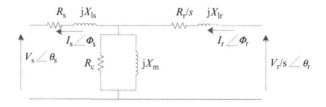

Figure 7.1 Steady-state DFIG circuit.

The last equation needs the knowledge regarding wind power's mechanical power versus rotating speed relationship. First of all, we need to express the mechanical power from the wind turbine by the unknown variables. Assume that the converters are lossless. Therefore, the power loss inside DFIG is due to the stator and rotor resistors. In addition, the gearbox loss can also be included.

$$P_{\text{gen,loss}} = I_s^2 R_s + I_r^2 R_r \tag{7.6}$$

$$P_{\text{gear,loss}} = \eta P_m + \xi P_{\text{rated}} \frac{\omega_m}{\omega_{m,\text{rated}}} \tag{7.7}$$

where $\eta = 0.02$ and $\xi = 0.005$ for a 2 MW gear box.

For simplicity, we can ignore the gearbox loss and express the mechanical power as $P_m = P_{\text{DFIG}} + I_s^2 R_s + I_r^2 R_r$.

Assume that the wind generator is operating at its maximum power extracting mode, then the relationship between P_m and slip is (refer Section 2.3)

$$P_m = k(1 - s)^3 \tag{7.8}$$

In the end, we are able to find the last equation $P_{\text{DFIG}} + I_s^2 R_s + I_r^2 R_r - k(1 - s)^3 = 0$.

To summarize, there are seven equations related to the seven unknown variables. These seven equations are

$$f_1 = V_s \cos \theta_s + R_s I_s \cos \phi_s - (X_{ls} + X_m) I_s \sin \phi_s + X_m I_r \sin \phi_r = 0 \tag{7.9}$$

$$f_2 = V_s \sin \theta_s - R_s I_s \sin \phi_s + (X_{ls} + X_m) I_s \cos \phi_s - X_m I_r \sin \phi_r = 0 \tag{7.10}$$

$$f_3 = V_r \cos \theta_r - s X_m I_s \sin \phi_s - R_r I_r \cos \phi_r + s(X_{lr} + X_m) I_r \sin \phi_r = 0 \tag{7.11}$$

$$f_4 = V_r \sin \theta_r - s X_m I_s \cos \phi_s - R_r I_r \sin \phi_r - s(X_{lr} + X_m) I_r \cos \phi_r = 0 \tag{7.12}$$

$$f_5 = P_{\text{DFIG}} - V_s I_s \cos(\theta_s - \phi_s) + V_r I_r \cos(\theta_r - \phi_r) = 0 \tag{7.13}$$

$$f_6 = Q_s - V_s I_s \sin(\theta_s - \phi_s) \tag{7.14}$$

$$f_7 = P_{\text{DFIG}} + I_s^2 R_s + I_r^2 R_r - k(1 - s)^3 = 0 \tag{7.15}$$

Let $X = \begin{bmatrix} V_r & \theta_r & I_s & \phi_s & I_r & \phi_r \end{bmatrix}^T$ and $f = \begin{bmatrix} f_1 & f_2 & f_3 & f_4 & f_5 & f_6 & f_7 \end{bmatrix}^T$. To solve the set of nonlinear equations, we can employ Newton-Raphson(NR) iterative method.

$$X^{k+1} = X^k - J^{-1}f \tag{7.16}$$

where $J = \frac{\partial f}{\partial X}$.

7.2 INTERCONNECTION OF DFIG MODEL IN POWER SYSTEMS

Modeling wind farms interconnected to the grid is important for the transient analysis of the entire system. In modeling a power system, the network is usually treated as a Y matrix with the current and voltage relationship as $\bar{I} = Y\bar{V}$, where $\bar{}$ refers to phasors. All generators are treated as current sources. From the current sources, the system voltages can be computed as in Fig. 7.2. The voltages will then be used in the differential equations expressed in terms of state variables $\bar{I}$ and input variables $\bar{V}$. In the case of a DFIG, the current to the network is the sum of the stator current and the current from the converter and the network voltage is the stator voltage.

In a power system, the voltage phasors are all based on a reference bus while the rotor angles are all based on a reference machine. The stator voltage phasor of a DFIG can be expressed as

$$\bar{V}_s = \frac{v_{qs} - jv_{ds}}{\sqrt{2}} = |V_s|\angle\phi. \tag{7.17}$$

In this case, the q-axis coincides with the reference voltage direction while the d-axis is lagging the q-axis by $90°$.

To facilitate the vector control design for the DFIG, the q-axis has to be aligned with the stator voltage $\vec{v_s}$. Therefore, the reference frame of the

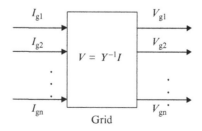

Figure 7.2 Grid representation.

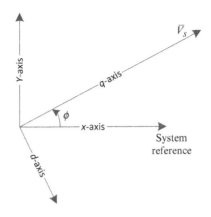

Figure 7.3 DFIG reference frame and the system reference frame.

Figure 7.4 Interface diagram.

system and the reference frame of the DFIG have an angle ϕ between them as in Fig. 7.3.

If the current to the network is in the machine reference frame, it has to be transformed back to the system reference frame while interconnecting the DFIG to the power system.

The modeling diagram from the stator voltage and the total current from the DFIG to the grid is shown in Fig. 7.4.

7.3 APPLICATION OF THE MODEL: INTERAREA-OSCILLATION DAMPING THROUGH DFIG WIND TURBINES

The control problem is stated as follows. The power system has an inherent issue of inter-area oscillations. With the deployment of Type-3 wind farms where power electronic converters are equipped, we would like to explore the capability of fast power modulation by the converters to improve the damping of inter-area oscillations. This example was also documented in [5].

Our approach is to design an auxiliary damping controller on top of the existing rotor converter control. To facilitate the design, we need to first have the plant model or the open-loop system model available. We will then based on the plant model design the feedback control system. The auxiliary damping controller's output will be a power command that modulates the power reference of the rotor side converter. The input of the controller will be a signal that has sufficient information of the oscillation. Angle difference between two areas will be a suitable signal. Therefore, the plant model has an input of power command and an output of the angle difference. In the following section, we first examine the study system and the existing RSC converter control. Then the plant model Bode plots are given. Based on the Bode plots, a reduced-order plant model is derived for controller design. Root loci-based control design is then conducted. The designed controller is tested in nonlinear dynamic simulation to check its feasibility in oscillation damping.

7.3.1 Configuration of the Test System

The test system shown in Fig. 7.5 is based on the classic two-area four-machine system developed in [6] for inter-area oscillation analysis. Area 1 has two synchronous generators, each with 835 MW rated power and Area 2 also has two synchronous generators, each with 835 MW rated power. All four synchronous generators are identical. In this paper, the full-order model of the synchronous generators is used. The parameters of the steam turbine generators come from Krause's classic textbook *Analysis of Electric Machinery* [7]. The parameters are also shown in Appendix. In Area 1, a wind farm is connected to the grid. The wind farm penetration is assumed to be 25% of Area 1, i.e., the rated exporting power level of the wind farm is 200 MW.

The transfer level along the tie line of the two areas varies from 0 to 400 MW due to the variation of load levels in the two areas. The inter-area oscillation is characterized as the swing of the generators in Area 1 against the generators in Area 2 and the frequency of the oscillation is about 0.7 Hz or 4.35 rad/s. In addition, there are local oscillations present in the system at about 7.7 rad/s. The local oscillations are characterized as the swing of one synchronous generator against the other in the same area.

7.3.2 Plant Model

The linearized plant model will be obtained through the dynamic system model. The entire power system dynamic model is first built in

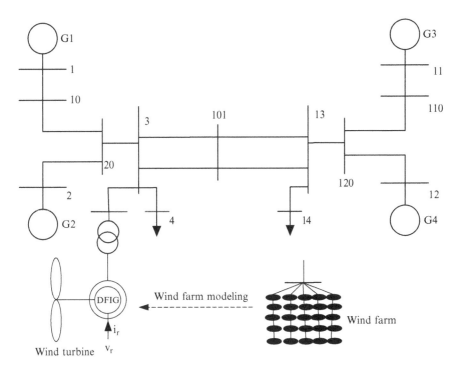

Figure 7.5 One-line diagram of two-area system with wind generator.

Matlab/Simulink.This dynamic model will be tested to have flat runs giving the initial state variables are accurate from initialization or steady-state calculation. With the input port defined as the RSC control power command and the output port as the difference of two rotor angles (Generators 3 and 1), we can obtain a linearized model at the current operating condition using Matlab's function "linmod." From this model, Bode plots will be obtained.

The overall control diagram for a rotor-side converter (RSC) is presented in Fig. 7.6. The details of each control block can be referred to Chapter 3 Converter control. Federal Energy Regulatory Commission (FERC) Order 661 requires the wind power plants to have the capability to control their reactive power so that the power factor falls within 0.95 leading to 0.95 lagging range [8]. Therefore the capability to track PQ is a basic requirement for a DFIG system. Active power and reactive power (PQ) tracking can be achieved by two PI controllers as shown in Fig. 7.6.

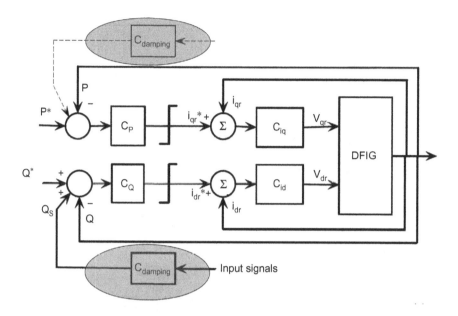

Figure 7.6 The overall control structure of the DFIG with rotor side converter.

7.3.3 Damping Control of DFIG

Since the current loop is very fast and its bandwidth is very high compared to the desired bandwidth of the damping control loop, there is no need to put a supplementary signal at the current control loop. Instead, we propose to add a supplementary signal at the active power control loop. Since the inter-area oscillation is a phenomenon related to the rotor angle and active power, active power modulation is an effective method for damping oscillations in power systems.

The rotor angle difference has a good observability of the inter-area oscillation mode between the two areas [9]. The rotor angle difference can be estimated from local measurements (voltage and current) using the method of [9]. The angle difference signal can be obtained accurately using a state-of-art Phasor Measurement Unit. In this example, we assume that the angle difference signal is available through either phasor measurement technology or local measurement and estimation.

The open-loop frequency responses of two systems, one with P modulation and a second with Q modulation are compared in Fig. 7.7. The first system relates the P modulation and the rotor angle difference and the second system relates the Q modulation and the rotor angle difference. It is

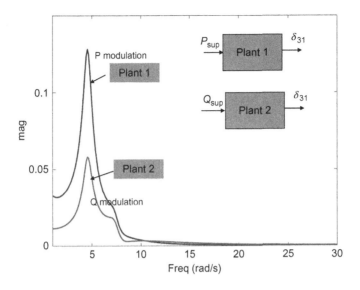

Figure 7.7 Open-loop frequency responses with P and Q modulations.

seen that the oscillation frequency is about 4 rad/s at which the first system has a higher magnitude compared to the second. To control the first system, we will need a smaller gain which is preferred since it avoids controller saturation. Thus, the frequency responses confirm that the active power modulation is a good choice for damping inter-area oscillations.

The root locus diagram of the open loop system is shown in Fig. 7.8. The open loop system is unstable since there are two pairs of complex poles on the RHP. One pair of poles corresponds to the inter-area oscillation mode at 4.73 rad/s and the other pair corresponds to an oscillation mode at 7.74 rad/s. The later oscillation mode is a local oscillation mode characterized by one generator oscillating against the other in the same area. The second pair of poles has the associated zeros close-by on the LHP. Therefore, it is impossible to move the second pair of poles to further left. The best option is to move the closed-loop poles as close to the zeros as possible.

A simple proportional controller cannot do the job due to the fact that the two oscillation modes (root locus) will move in opposite directions according to the root locus diagram. Moving one mode to the left plane means moving the other mode to the right plane. The order of the open-loop system is high so a more complicated controller is required. The design is

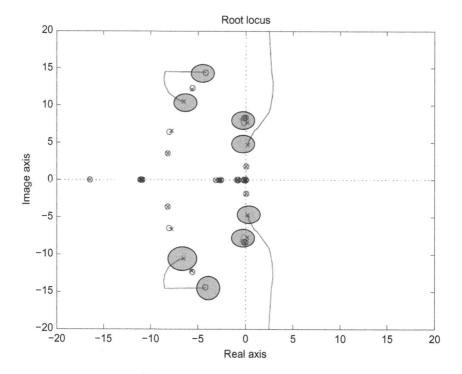

Figure 7.8 Root locus diagram for the open loop system.

simplified by considering the dominant zeros and poles only (highlighted in Fig. 7.8). The transfer function of the approximate system is given by

$$P = \frac{(s + 0.177 \pm j8.32)(s + 4.22 \pm j14.5)}{(s - 0.209 \pm j4.79)(s - 0.15 \pm j7.67)(s + 6.5 \pm j10.5)}. \quad (7.18)$$

The frequency response of the approximate system P is compared with that of the original system (Fig. 7.9). It is seen that the phase angles are the same over the frequency range 1-100 rad/s. However, there are magnitude differences between the two. Compensating P with a gain $k = 0.56$ will make the frequency responses of Pk and the original system the same as shown in Fig. 7.9.

For the approximate lower order system Pk, we now design a controller that can move the two pairs of unstable poles to the left plane. It is found that the first pair of poles corresponds to the inter-area oscillation mode which has a frequency of 4.73 rad/s. The second pair of the poles corresponds to an oscillation mode with a frequency of 7.74 rad/s. Apparently, the selected

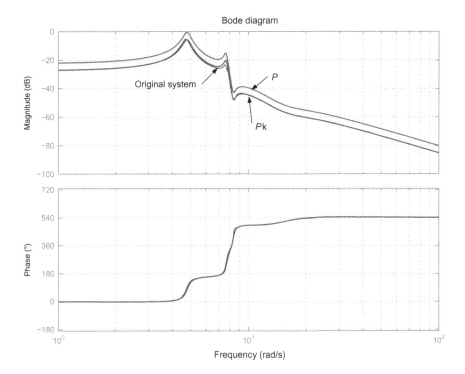

Figure 7.9 Bode plots of the original system and the approximate system.

input signal (rotor angle difference) is not effective in damping the second oscillation mode since the poles are very close to the zeros. To enhance the damping of the inter-area oscillation mode, a pair of complex zeros ($-0.05\pm$ $j0.77$, notated as 1 in Fig. 7.10) is added in the left plane close to the poles. These zeros cause the closed-loop poles move to the left plane when the gain increases. To make the controller proper, two real poles (-6.7,-62.5) on the left real axis are also added. To make the second pair of closed-loop poles move to the corresponding zeros as fast as possible, two pairs of zeros-poles ($-0.62\pm j6.64$- notated as 2 and $-1.24\pm j9.01$- notated as 3 in Fig. 7.10) nearby are added. The controller designed (G_d) is given by

$$G_d = \frac{(1 + 0.17s + 1.3^2 s^2)(1 + 0.028s + 0.15^2 s^2)}{(1 + 0.15s)(1 + 0.016s)(1 + 0.03s + 0.11^2 s^2)}. \tag{7.19}$$

The root locus diagram of the compensated system kG_dP is also shown in Fig. 7.10.

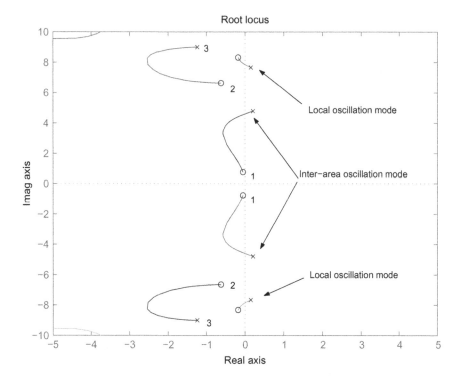

Figure 7.10 Root locus diagram after compensation.

The gain that corresponds to maximum damping of the inter-area oscillation mode is computed from the root locus diagram as 2.0. The final damping controller designed is given by

$$C_{\text{damping}} = 2\frac{(1 + 0.17s + 1.3^2s^2)(1 + 0.028s + 0.15^2s^2)}{(1 + 0.15s)(1 + 0.016s)(1 + 0.03s + 0.11^2s^2)}. \quad (7.20)$$

7.3.4 Validation Through Time-Domain Simulation

Time-domain Simulations are performed on the test system in order to show the effectiveness of the damping control. The system operates under steady state for 0.1 s with the power transfer between the two areas at 400 MW. A temporary three-phase fault occurs at Bus 3 at 0.1 s and is cleared subsequently at 0.1 s. Figures 7.11 and 7.12 show the dynamic responses of the synchronous generators and the DFIG when there is no inter-area oscillation control. In Fig. 7.11, the relative angle differences, rotor speeds and electric power export levels are plotted. In Fig. 7.12, the rotor speed, mechanical torque, electric torque and terminal voltage of the DFIG are

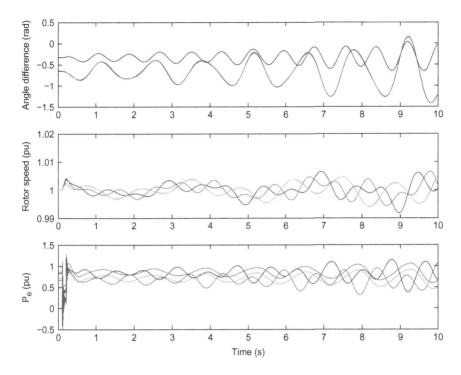

Figure 7.11 Synchronous generator dynamic responses with no supplementary damping control. (a) δ_{21}, δ_{31} and δ_{41}, (b) rotor speeds of four generators, (c) output power from generators.

plotted. It is seen that the system exhibits low-frequency oscillations whose amplitude increases with time and eventually the system becomes unstable.

Figures 7.13–7.15 show the dynamic responses of the synchronous generators, the induction generator, and DFIG damping controller with the auxiliary damping control added for active power modulation.

The dynamic responses of the rotor angles are plotted together for the two scenarios: (1) with no damping control, (2) with damping control. The plots are shown in Fig. 7.16 where dashed lines correspond to the first scenario and the solid lines correspond to the second scenario. From the plots, we can observe the effectiveness of the designed controller in damping out the 0.7 Hz inter-area oscillation. Meanwhile the other oscillation mode tends to damp out as well. Compared to the originally unstable case, the modified system becomes stable with the addition of a supplementary control loop. The enhanced stability can further help to improve transfer capability and move more wind power to the market.

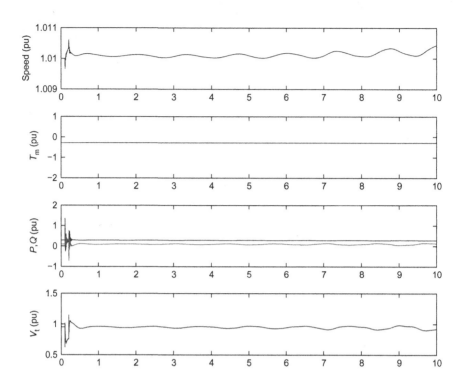

Figure 7.12 Wind turbine generation dynamic responses with no supplementary damping control: (a) speed of DFIG rotor; (b) mechanical torque; (c) output P and Q of the DFIG (P is above curve Q); (d) terminal voltage magnitude of the DFIG.

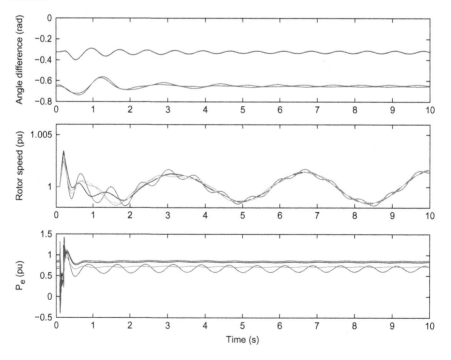

Figure 7.13 Synchronous generator dynamic responses with supplementary damping control: (a) δ_{21}, δ_{31}, and δ_{41}; (b) rotor speeds of four generators; (c) output power from generators.

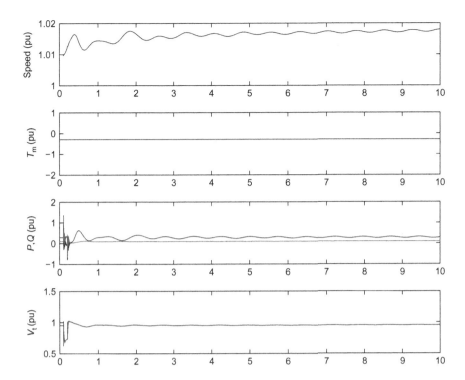

Figure 7.14 Wind turbine generator dynamic responses with supplementary damping control: (a) speed of DFIG rotor; (b) mechanical torque; (c) output P and Q of the DFIG (P is above curve Q); (d) terminal voltage magnitude of the DFIG.

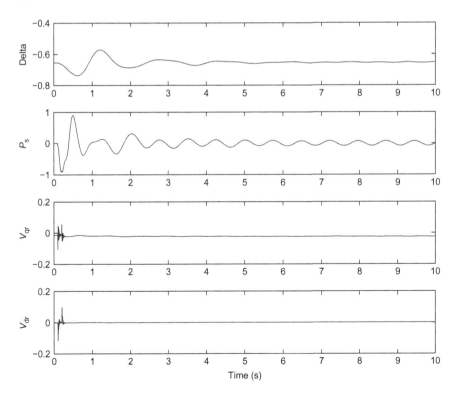

Figure 7.15 Damping control input, output signals and v_{qr} and v_{dr} dynamic responses.

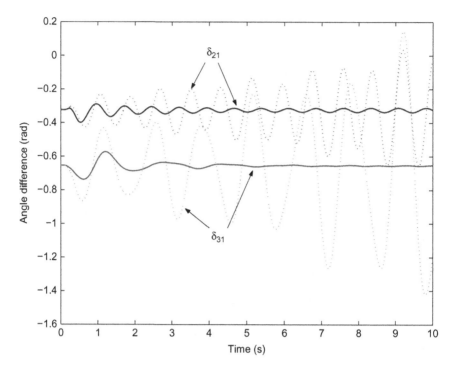

Figure 7.16 Rotor angle dynamic response comparison (solid lines correspond to the dynamic responses for the system with damping control; dotted lines correspond to the dynamic responses for the system without damping control).

This example has demonstrated the use of the multi-machine power system model (with Type-3 wind generator) for inter-area oscillation damping control.

APPENDIX

Synchronous generator parameters:

Rating: 835 MVA, line to line voltage: 26 kV, poles: 2, speed: 3600 r/min
Combined inertia of generator and turbine: $H = 5.6$ s
$r_s = 0.00243 \, \Omega, 0.003 \, \text{pu}, X_{ls} = 0.1538 \, \Omega, 0.19 \, \text{pu}$
$X_q = 1.457 \, \Omega, 1.8 \, \text{pu}, X_d = 1.457 \, \Omega, 1.8 \, \text{pu}$
$r'_{kq1} = 0.00144 \, \Omega, 0.00178 \, \text{pu}, r'_{fd} = 0.00075 \, \Omega, 0.000929 \, \text{pu}$
$X'_{lkq1} = 0.6578 \, \Omega, 0.8125 \, \text{pu}, X'_{lfd} = 0.1165 \, \Omega, 0.1414 \, \text{pu}$
$r'_{kq2} = 0.00681 \, \Omega, 0.00841 \, \text{pu}, r'_{kd} = 0.01080 \, \Omega, 0.01334 \, \text{pu}$
$X'_{lkq2} = 0.07602 \, \Omega, 0.0939 \, \text{pu}, X'_{lkd} = 0.06577 \, \Omega, 0.08125 \, \text{pu}$

Induction generator parameters:

$$H = 5\text{s}, \; r_\text{s} = 0.00059\,pu, \; X_M = 0.4161\,pu, \; r_\text{r} = 0.00339\,pu,$$
$$X_\text{ls} = 0.0135\,pu, X_\text{lr} = 0.0075\,pu.$$

Current control loops: $0.0352 + \frac{1.6765}{s}$.

PQ control loops: $1 + \frac{1}{s}$.

REFERENCES

[1] P. Kundur, N.J. Balu, M.G. Lauby, Power System Stability and Control, vol. 7, McGraw-Hill, New York, 1994, vol. 7.

[2] P.W. Sauer, M. Pai, Power System Dynamics and Stability, vol. 4, Prentice Hall, Upper Saddle River, NJ, 1998.

[3] Arthur R. Bergen, V. Vittal, Power systems analysis. Prentice Hall, 1999.

[4] Miao, Zhixin, and Lingling Fan. "The art of modeling and simulation of induction generator in wind generation applications using high-order model." Simulation Modelling Practice and Theory 16.9 (2008) 1239-1253.

[5] Z. Miao, L. Fan, D. Osborn, S. Yuvarajan, Control of dfig-based wind generation to improve interarea oscillation damping, IEEE Trans. Energy Convers. 24(2) (2009) 415-422.

[6] M. Klein, G. Rogers, P. Kundur, A fundamental study of inter-area oscillations in power systems, IEEE Trans. Power Syst. 6 (1991) 914-921.

[7] P. Krause, Analysis of Electric Machinery, McGraw-Hill, New York, 1986.

[8] R. Zavadil, N. Miller, A. Ellis, E. Muljadi, Queuing up, IEEE Power Energy Mag. 5(6) (2007) 47-58.

[9] E. Larsen, J.J. Sanchez-Gasca, J. Chow, Concepts for design of FACTS controllers to damp power swings, IEEE Trans. Power Syst. 10 (1995) 948-956.

Printed in the United States
By Bookmasters